Sustainable Production, Life Cycle Engineering and Management

Series Editors

Christoph Herrmann, Braunschweig, Germany

Sami Kara, Sydney, Australia

SPLCEM publishes authored conference proceedings, contributed volumes and authored monographs that present cutting-edge research information as well as new perspectives on classical fields, while maintaining Springer's high standards of excellence, the content is peer reviewed. This series focuses on the issues and latest developments towards sustainability in production based on life cycle thinking. Modern production enables a high standard of living worldwide through products and services. Global responsibility requires a comprehensive integration of sustainable development fostered by new paradigms, innovative technologies, methods and tools as well as business models. Minimizing material and energy usage, adapting material and energy flows to better fit natural process capacities, and changing consumption behaviour are important aspects of future production. A life cycle perspective and an integrated economic, ecological and social evaluation are essential requirements in management and engineering.

Indexed in Scopus

To submit a proposal or request further information, please use the PDF Proposal Form or contact directly: Petra Jantzen, Applied Sciences Editorial, email:petra.jantzen@springer.com

More information about this series at http://www.springer.com/series/10615

Felipe Cerdas

Integrated Computational Life Cycle Engineering for Traction Batteries

Springer

Felipe Cerdas
Institute of Machine Tools and Production Technology
Technische Universität Braunschweig
Braunschweig, Germany

ISSN 2194-0541 ISSN 2194-055X (electronic)
Sustainable Production, Life Cycle Engineering and Management
ISBN 978-3-030-82933-9 ISBN 978-3-030-82934-6 (eBook)
https://doi.org/10.1007/978-3-030-82934-6

This Springer imprint is published by the registered company Springer Nature Switzerland AG
The registered company address is: Gewerbestrasse 11, 6330 Cham, Switzerland

Acknowledgements

This book is the result of my work as a research associate at the Institute of Machine Tools and Production Technologies (IWF) of the Technische Universität Braunschweig. During this time, I had the great opportunity to collaborate and manage several research projects, which have, one way or another, provided a basis for the development of the concept presented in this book.

In this regard, I want to express my gratitude to the German Academic Exchange Service (DAAD) for awarding me with the scholarship that supported the beginning of this journey. Thank you very much also to the Federal Ministry of the Environment, Nature Conservation and Nuclear Safety (BMU) for the funding provided in the context of the project LithoRec II and to the Federal Ministry of Education and Research (BMBF) for the funding provided for the project Benchbatt.

Many, many thanks to my supervisor Prof. Dr.-Ing. Christoph Herrmann, Head of the Institute of Machine Tools and Production Technology. Christoph gave me a creative working environment, all freedom, trust and resources to explore new ideas and a pragmatic perspective that helped me to better understand the research needs in engineering towards sustainability.

Special thanks to Prof. Dr.-Ing. Jens Friedrichs, Head of the Institute of Jet Propulsion and Turbomachinery (IFAS), for his engagement as chair of the doctoral dissertation committee. Thank you very much also to my co-examiners Prof. Dr.-Ing. Arno Kwade, Head of the Institute for Particle Technology, and to Prof. Dr. Anders Hammer Strømman, professor within the Industrial Ecology Program at NTNU, for their tough questions and very useful feedback and comments.

Working at IWF has been a wonderful experience, and really each of the team members I have worked with throughout these years has made me feel welcomed and happy working here. Special thanks to Sebastian Thiede, Tina Dettmer and Mark Mennenga for all their guidance and constant motivation. I would like to express my gratitude to each of the students of whom I had the honor to supervise her/his thesis. They have contributed greatly to my productive and happy days at the institute.

Finally, I want to express my most deep and sincere gratitude to all my friends and members of my family. Special thanks to my parents for their love and unconditional support. Finally, million thanks to my wife for all her patience and tireless encouragement all these years, for always reminding me that there is a life besides research and for helping me in all possible ways to bring this phase to a successful end.

Braunschweig, Germany
2021

Felipe Cerdas

About This Book

Modelling, analysing and understanding the environmental impacts of a product system over its life cycle is a time-consuming and interdisciplinary task. It requires specialized life cycle assessment (LCA) modelling skills and deep knowledge of the product system under analysis. EVs are particularly complex product systems. Their environmental impacts are sensitive to a large set of parameters in the foreground system, and they interact extensively with temporal, geographical and technical contexts. Current LCA software tools do not enable an efficient integration of this complexity into analytical models, leading to the development of simplified tools that hamper well-informed decision-making processes and hold back the implementation of LCA within engineering activities. In addition, current approaches limit the application of visual analytics due to the inadequacies of the results achieved. Consequently, while there is an increasing number of qualitative LCA studies on traction batteries, comparable LCA research in the field electromobility is unfortunately fraught with diverging results, rising from specific subjective modelling choices. To face these challenges, an Integrated Computational Life Cycle Engineering (ICLCE) framework for EVs has been developed in the context of this work. This ICLCE framework aims to support a fast and comprehensive modelling of complex foreground systems in the electromobility field and to show their interaction with diverse backgrounds and spatial contexts.

The ICLCE concept is envisioned as a plug and play-like platform that enables coupling and freely exchanging multidisciplinary models in different scales for all life cycle stages of a product system. This flexible integration of multidisciplinary models enables the variability of parameters and expert knowledge to be considered, resulting in a more robust support for engineering decisions. The concept is implemented as a software prototype; namely, it is a platform to integrate discipline-specific models representing distinct unit processes in the life cycle of a traction battery. The ICLCE framework presents a complete exemplary application process, beginning with setting the research question and defining the system boundaries, continuing with documenting the models and model coupling strategy, and concluding with a presentation and analysis of the results.

Contents

Acronyms

AoP	Areas of protection
BEV	Battery electric vehicle
BMS	Battery management system
BMU	Battery management unit
BS	Background system
CADC	Common Artemis Driving Cycle
CS	Computational system
DoH	Degree of hybridization
EPA	Environmental Protection Agency
EREV	Extended range electric vehicle
EV	Electric vehicle
FS	Foreground system
GHG	Greenhouse gas
GUI	Graphical user interface
GWP	Global warming potential
HEV	Hybrid electric vehicle
HVAC	Heating, ventilation and air conditioning
ICE	Internal combustion engine
ICME	Integrated Computational Materials Engineering
IEA	International Energy Agency
LCA	Life cycle assessment
LCE	Life Cycle Engineering
LCI	Life cycle inventory
LCIA	Life cycle impact assessment
LIB	Lithium-ion battery
MBSE	Model-Based Systems Engineering
NEDC	New European Driving Cycle
PHEV	Plug-in hybrid electric vehicle
PMSM	Permanent magnet synchronous machine
PO	Product object
PS	Product system

SEI	Solid electrolyte interphase
SoC	State of charge
SoH	State of health
TTW	Tank-to-wheel
WLTP	Worldwide Harmonized Light Vehicles Test Procedure
WTT	Well-to-tank
WTW	Well-to-wheel

Equations

2.1 $I = \mathrm{P} \cdot \mathrm{A} \cdot \mathrm{T}$

2.2 $DoH = \frac{P_{b,max}}{P_{b,max} + P_{e,max}}$

2.3 $m_v \frac{d}{dt} v(t) = F_t(t) - F_a(t) + F_r(t) + F_g(t) + F_d(t)$

2.4 $F_a = \frac{1}{2} \cdot \rho_a \cdot A_f \cdot c_d \cdot v^2$

2.5 $F_r = c_r \cdot m \cdot g \cdot \cos(\alpha)$

2.6 $F_g = m \cdot g \cdot \cos(\alpha)$

2.7 $F_k = m_e \cdot a$

2.8 $P_w = F_t \cdot v$

2.9 $P_{bat,out} = \frac{P_w}{\gamma_{bat,disc} \cdot \gamma_{pe} \cdot \gamma_{em} \cdot \gamma_t}$

2.10 $E_m = \frac{1}{x_{tot}} \int_{t=0}^{T} P_{bat,out} dt$

5.1 $c_{vol|am} = q_{am} \cdot \rho_{am|crystal}$

5.2 $\rho_{com} = \frac{1}{\frac{mf_{am}}{\rho_{am|crystal}} + \frac{mf_{bin}}{\rho_{bin}} + \frac{mf_{ca}}{\rho_{ca}}}$

5.3 $\rho_{el|am} = mf_{am} \cdot \rho_{com}(1 - \varepsilon)$

5.4 $\rho_{el} = (1 - \varepsilon) \cdot \rho_{com} + \varepsilon \cdot \rho_{elek}$

5.5 $c_{vol|el} = q_{am} \cdot \rho_{el|am}$

5.6 $E_{mat|grav} = q_{am} \cdot V_{am \mid vs\ Li/Li}$

5.7 $E_{mat|vol} = c_{vol|el} \cdot V_{am \mid vs\ Li/Li}$

5.8 $Am_{pos} = t_{pos} \cdot \rho_{pos|am}$

5.9 $c_{area|pos} = t_{pos} \cdot c_{vol|pos}$

5.10 $c_{area|neg} = c_{area|pos} \cdot np_{ratio}$

5.11 $t_{neg} = \frac{c_{area|neg}}{c_{vol|neg}}$

5.12 $Am_{neg} = t_{neg} \cdot \rho_{neg|am}$

5.13 $t_{stack} = t_{pos} + t_{neg} + \frac{t_{pos|coll}}{2} + \frac{t_{neg|coll}}{2} + t_{sep}$

5.14 $V_{nom} = V_{pos \mid vs\ Li/Li} - V_{pos \mid vs\ Li/Li}$

5.15 $E_{stack|vol} = V_{nom} \cdot \frac{c_{area|pos}}{t_{stack}}$

5.16 $E_{stack|grav} = V_{nom} \cdot \frac{c_{area|pos}}{Am_{stack}}$

5.17 $n_{stack} = \frac{t_{cell}}{t_{stack}}$

5.18 $E_{trac} = \frac{1}{x_{tot}} \int_{P_w > 0} (P_a + P_r + P_k) dt$

5.19 $E_{trac} = \frac{1}{x_{tot}} \int\limits_{P_w>0} \left(\frac{1}{2} \cdot \rho \cdot c_d \cdot A_f \cdot v^3 + c_r \cdot m \cdot g \cdot v + m \cdot a \cdot v\right) dt$

5.20 $E_{trac} = A \cdot c_d \cdot A_f + B \cdot c_r \cdot m + C \cdot m$

5.21 $A = \frac{\rho}{2 \cdot x_{tot}} \int\limits_{P_w>0} v^3 dt$

5.22 $B = \frac{g}{x_{tot}} \int\limits_{P_w>0} v\, dt$

5.23 $C = \frac{1}{x_{tot}} \int\limits_{P_w>0} a \cdot v\, dt$

5.24 $E_{regen} = A' \cdot c_d \cdot A_f + B' \cdot c_r \cdot m + C' \cdot m$

5.25 $E_{diss} = \left(A + A'\right) \cdot c_d \cdot A_f + \left(B + B'\right) \cdot c_r \cdot m$

Symbols

I	Impact (-)
P	Population (-)
A	Affluence (-)
T	Technology (-)
DoH	Degree of hybridization (dimensionless)
$P_{b,max}$	Battery's maximum capable power (kW)
$P_{e,max}$	Engine's maximum capable power (kW)
m_v	Vehicle's mass (Kg)
v	Velocity (m/s)
F_t	Traction force (N)
F_a	Aerodynamic force (N)
F_r	Rolling friction (N)
F_g	Gravity force (N)
F_d	Disturbance force (N)
ρ_a	Air density (kg/m^3)
A_f	Vehicle's frontal area (m^2)
c_d	Aerodynamic drag coefficient (dimensionless)
c_r	Rolling resistance coefficient (dimensionless)
F_k	Kinetic force (N)
m_e	Vehicle's equivalent mass (Kg)
a	Acceleration (m/s^2)
P_w	Power at the wheel
$P_{bat,out}$	Power provided by the battery at a given point (kW)
$\gamma_{bat,disc}$	Discharge efficiency of the battery (dimensionless)
γ_{pe}	Efficiency of the power electronics (dimensionless)
γ_{em}	Efficiency of the electric motor (dimensionless)
γ_t	Efficiency of the transmission (dimensionless)
E_m	Energy demand for a given driving cycle (kWh)
x_{tot}	Distance (km)
$c_{vol\|am}$	Volumetric capacity of the active material $\left[\frac{\text{mAh}}{\text{cm}^3}\right]$
q_{am}	Specific capacity of the material $\left[\frac{\text{mAh}}{\text{g}}\right]$

$\rho_{am\|crystal}$	Crystallographic density $\left[\frac{g}{cm^3}\right]$
ρ_{com}	Composite electrode density $\left[\frac{g}{cm^3}\right]$
mf_{am}	Mass fraction of active material (dimensionless)
mf_{bin}	Mass fraction of binder and conductive additive (dimensionless)
mf_{ca}	Mass fraction of binder and conductive additive (dimensionless)
ρ_{bin}	Density of the binder $\left[\frac{g}{cm^3}\right]$
ρ_{ca}	Density of conductive additive $\left[\frac{g}{cm^3}\right]$
$\rho_{el\|am}$	Density of the active material fraction in the electrode $\left[\frac{g}{cm^3}\right]$
ε	External porosity (dimensionless)
ρ_{el}	Electrode density including electrolyte $\left[\frac{g}{cm^3}\right]$
ρ_{elek}	Density of the electrolyte $\left[\frac{g}{cm^3}\right]$
$E_{mat\|grav}$	Specific material gravimetric energy $\left[\frac{Wh}{kg}\right]$
$E_{mat\|vol}$	Specific material volumetric energy $\left[\frac{Wh}{l}\right]$
$V_{am\ \|\ vs\ Li/Li}$	Average discharge potential [V]
Am_{pos}	Area mass of the positive electrode $\left[\frac{mg}{cm^2}\right]$
t_{pos}	Thickness of the positive active material coating (μ m)
$c_{area\|pos}$	Positive electrode area capacity $\left[\frac{mAh}{cm^2}\right]$
$c_{area\|neg}$	Negative electrode area capacity $\left[\frac{mAh}{cm^2}\right]$
np_{ratio}	Capacity ratio between the positive and negative electrodes (dimensionless)
t_{neg}	Thickness of the negative active material coating (μ m)
Am_{neg}	Area mass of the negative electrode $\left[\frac{mg}{cm^2}\right]$
t_{stack}	Total thickness of the stack (μ m)
$t_{pos\|coll}$	Thickness of the positive electrode current collector (μ m)
$t_{neg\|coll}$	Thickness of the negative electrode current collectors (μ m)
t_{sep}	Thickness of the separator (μ m)
Am_{stack}	Areal mass of the stack $\left[\frac{mg}{cm^2}\right]$
$V_{nom}[V]$	Nominal cell voltage [V]
$E_{stack\|vol}$	Stack volumetric energy $\left[\frac{Wh}{l}\right]$
$E_{stack\|grav}$	Stack gravimetric energy $\left[\frac{Wh}{kg}\right]$
E_{trac}	Traction energy (kWh)
E_{regen}	Regeneration energy (kWh)
E_{diss}	Dissipative energy (kWh)

Chapter 1
Background and Context

Check for updates

Contents

As the transport sector moves towards new technological alternatives that enable a significant reduction of greenhouse gases (GHG) produced during its operation stage, the complexity of the vehicles, their components, and supply chains increases. The current methodologies that quantify the potential environmental impact of these new technologies cannot effectively cope with this complexity, complicating the consideration of mitigation options within decision-making and engineering development activities. This chapter gives an overview of the current context of electromobility from an environmental perspective, while discussing for a change of paradigm in the application of current assessment methodologies. Finally, the outline and the context in which this research was developed are presented.

1.1 Towards a More Sustainable Private Transportation System

As opposed to public transportation, private transportation refers to the privately-owned vehicles which enable individual operators to freely decide the time, length, speed and route of their journey. Private transportation sector faces several challenges in achieving a sustainable development: This sector is responsible for 25% of urban air pollution from $PM_{2,5}$, creating serious human health damages (Karagulian et al. 2015). Additionally, the transportation sector produced 24% of all energy-related greenhouse gas (GHG) emissions in 2017, 75% of which were produced by road vehicles (International Energy Agency 2019a). Moreover, the road transportation

F. Cerdas, *Integrated Computational Life Cycle Engineering for Traction Batteries*, Sustainable Production, Life Cycle Engineering and Management,
https://doi.org/10.1007/978-3-030-82934-6_1

sector alone presented a 2% annual growth rate in the period from 2010 to 2017, and dynamic socio-economic developments in many emerging economies are expected to increase private transportation demand per capita, leading to the even faster growth of GHG emissions (Sims et al. 2014). According to the International Energy Agency (IEA), if concrete measures and actions to limit GHG emissions are not taken, GHG emissions from transport are expected to rise by 20% by 2030 and to around 50% by 2050 (International Energy Agency (IEA) 2016).

The IEA estimated that, in a 2 °C carbon abatement scenario, the transportation sector alone could potentially contribute to a reduction of 18% of GHG by 2050 (International Energy Agency 2019b). Numerous engineering and business initiatives have consequently appeared, aiming to reduce the impact of transportation. Some of these include (i) the development of alternative fuels, (ii) business models changing ownership and usage models, (iii) the improvement of vehicle's efficiency and (iv) the development of alternative power trains, as summarized in (Cerdas et al. 2018b).

In this context, the global deployment of electric vehicles (EVs) has emerged as an essential step to mitigate the environmental impacts of road transportation. The Paris Declaration on Electro-Mobility and Climate Change (UNFCCC 2015) set a global electric vehicles stock target of 20% (around 100 million vehicles), to be reached in 2030 to prevent an increase of the global temperature beyond 2 °C (UNFCCC 2015). Estimations from the IEA suggest a stock of over 140 million vehicles by 2030 in a scenario considering new policies developed worldwide (International Energy Agency 2019b) and up to 250 million vehicles in an scenario in which 30% of the planet's fleet is composed of EVs as shown in Fig. 1.1. While the current share of EVs in the global vehicle stock remains low (International Energy Agency 2019b), factors such as the fast decreasing battery prices (McKinsey 2017) and the increasing support from national policies have boost EV sales in developed regions like Europe, Japan, the United States and in emerging economies like China and India (International Energy Agency 2019b).

It has been widely suggested that EVs can make a significant impact in reducing the amount of GHG originating from the transport sector. This is because Evs do not produce exhaust emissions from combustion. Furthermore, their power train efficiency ranges between 60 and 80% (Sims et al. 2014) in contrast to a 20–35% power train efficiency for internal combustion engine (ICE) vehicles (Thomas 2014). What's more, if powered with energy with a low share of fossil fuels, Evs have the potential to significantly reduce GHG emissions from a well-to-wheel (WTW[1]) perspective. As estimated by the International Council on Clean Transportation (ICCT), operating an EV with energy from the European electricity grid mix can contribute to reducing over 37% more GHG emissions than a diesel ICE vehicle and around 46% more than a gasoline ICE vehicle (Wolfram and Lutsey 2016). Accordingly, as countries commit

[1] A well-to-wheel (WTW) analysis considers all direct and indirect emissions throughout the life cycle of an energy carrier (e.g. fuel, electricity, etc.) from its production to its consumption in form of kinetic energy at the vehicle's wheels. Alternative analyses are: i. tank-to-wheel (TTW), which considers the conversion from energy carriers into the kinetic energy required to move the vehicle, and ii. well-to-tank (WTT), which considers exclusively the production of the energy carriers (fuels, electricity, etc.).

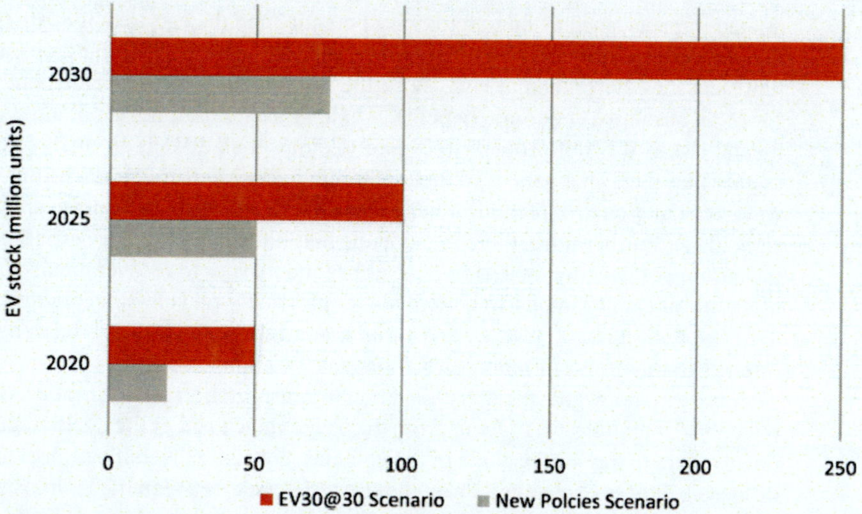

Fig. 1.1 Prediction of increase in electric vehicle stock and new registrations worldwide from (International Energy Agency 2019b). Based on IEA data from IEA (2019a, b) Global EV Outlook 2019, IEA (2019a, b), All rights reserved; as modified by Cerdas. www.iea.org/statistics, All rights reserved; as modified by [your legal entity name]

to increasing the share of energies from renewable sources in the electricity grid, WTW GHG emissions from Evs are expected to decrease drastically. Additionally, Evs can potentially help reduce local air and noise pollution in high exposure areas, a highly relevant issue as cities worldwide continue to grow steadily (International Energy Agency 2019b).

1.2 The Risk of Problem Shifting

A widespread adoption of EVs brings, however, new environmental and socioeconomic challenges. The shift in drivetrain technology implies changes in the material supply chains, mainly driven by the production of the battery system. Traction batteries are heavy, expensive, and short-lived devices, and are composed of environmentally intensive materials such as aluminum, copper, nickel (Dunn et al. 2015; Nelson et al. 2017). Some of these are linked to conflict minerals supply chains (e.g. cobalt, lithium) due to unstable geopolitical contexts or social risks related to their production (Schmidt et al. 2016; Gemechu et al. 2017). The battery system of an EV has been estimated to contribute to 35–44% of the overall Global Warming Potential (GWP) (Hawkins et al. 2013; Volkswagen AG 2013; Ellingsen et al. 2014; Dunn et al. 2015; Cerdas et al. 2018c) caused during its production stage. Additionally, the provision of this device has been linked to other important toxicity related impacts, such as human toxicity potential and the eco-toxicity caused by the mining tailings, among other factors. Meanwhile, spent traction batteries may also represent an environmental risk. With the absence of waste management policies (or of an attractive economic benefit), a battery system implies a potential environmental risk if it is not properly recycled (Cerdas et al. 2018a) as small fractions of metals like zinc, manganese, copper, lithium and nickel might leach into soil and water, causing environmental issues to health and ecosystems (European Commission 2014). Furthermore, there is evidence that informal recycling activities such as the burning of isolation materials like rubber and plastics to recover metals, or the use of environmentally intensive chemicals to informally recycle valuable metals, pose a high pollution risk to soil, air and water that might consequently lead to dangerous impacts to the health of humans (Kahhat and Williams 2012).

A transition towards a battery-dependent transportation sector implies, therefore, not only a potential shift of focus from the usage stage to other life cycle stages up and downstream of the supply chain of the vehicles, but also a diversification of the nature of the substances exchanged throughout its life cycle between the technosphere and the biosphere, and thus the respective damage to the environment. In this regard, the system becomes, from both an engineering and environmental perspective, more complex to assess and construct.

1.3 Complex Systems Require Tools and Methodologies

Engineering's most important task is to deal with the complexity of systems. In order to do so, methods and tools to effectively grasp complexity and to design systems are needed (Velten 2008; Herrmann 2010). Life Cycle Assessment (LCA) has in the last 20 years emerged as a system analysis methodology that enables a science-based quantification of the environmental impacts of product systems using a life cycle perspective on a set of environmental impact categories (EC-JRC 2010; Hellweg and Mila i Canals 2014). LCA can be applied at different stages of decision making, such as a support for policy making, as well as during engineering activities like product development (Hauschild et al. 2017). Nevertheless, as the complexity of the product increases, the methodology and tools enabling its application reach their limits, becoming highly time and resource-intensive (Baitz et al. 2013). This has made LCA unfit to cover the whole spectrum of results when considering a higher variability in the configuration of the product system and thus is inadequate in regard to other important influencing factors. This is problematic for two reasons. First, through the simplification of the system (i.e. by not considering all relevant influencing factors, e.g. technological, geographical, dimensional and cultural), LCA results might mislead strategic decision-making processes. Second, it becomes impossible to generate enough useful environmental insights that can be used as engineering constraints within product or process development activities. More problematic is the fact that, due to the inherent modelling freedom of LCA, it can be used to make misleading environmental claims about a system under the guise of having applied scientific methodology. This has not only hampered the integration of LCA within engineering activities such as product and process development, but it has sometimes led to disputed and controversial studies erroneously making assertive comparison between different systems and granting advantages to specific technologies on the basis of a subjectively defined system boundary.

More than serving as an emissions accounting instrument, LCA has the potential of disclosing important environmental insights that can be used to increase the resolution needed in the customization of a system, leading to more environmentally consistent designs. In other words, if engineers change the focus from a value and assessment-oriented application (resulting, for instance, in a value e.g. 120 g CO_2-eq/km) to a variability and uncertainty-oriented application (resulting in a model, a range, a constraint, etc.), more reliable solutions can be developed and better optimization strategies can be found. This is perhaps one of the most significant benefits of embedding LCA and its scientific foundation within engineering activities.

This is particularly relevant when analyzing the transformation in the transportation sector. While the environmental impact of ICEs has been predominantly located in the usage stage, and defined as the consumption of fossil fuels and the generation of tailpipe emissions, the improvement strategies have been centered mostly on increasing TTW efficiency, a task that automotive engineers have mastered. In contrast, the life cycle of an EV is very complex and largely unconstrained, due in part to the fact that the traction battery required for an EV is still heavily under research (Kwade et al. 2018; Michaelis et al. 2018; Schmuch et al. 2018). Additionally, the production and supply chains of key materials are widely different from each other depending on the source of origin, the use phase is highly sensitive to the local context where the vehicle is driven and, most importantly, a battery system is not in fact a "battery system". Rather, the technology development roadmap is still quite open, with researchers and manufacturers striving to optimize these products in terms of costs and technical properties, a wide range of battery materials, cell geometries, and manufacturing processes, to mention just a few of the considerations researchers are working with. As seen in Fig. 1.2, improving key properties of battery cells, such as their gravimetric energy, is linked to both the development and optimization of new materials and the manufacturing processes required to produce them.

Assessing the environmental impact of EVs, though, requires taking a life cycle perspective and the consideration of a substantial degree of variability in the configuration of the system and its interaction with other systems. As a consequence, current knowledge regarding the life cycle environmental impacts of the batteries used in EVs strongly varies among the studies currently available in the scientific literature. This is shown in Fig. 1.3 with results varying from 35 kg CO_2-eq/kWh up to 350 kg CO_2-eq/kWh.

This huge difference in peer-reviewed published results corresponds with the fact that the assumptions and modelling choices made while performing the LCA also differ among the studies. It does not necessarily imply that any given study is wrong, but rather that each study is the result of a particular LCA for the analysis of a particular traction battery system related to a specific set of external factors. Engineering solutions attempting to reduce the environmental impact of EVs using

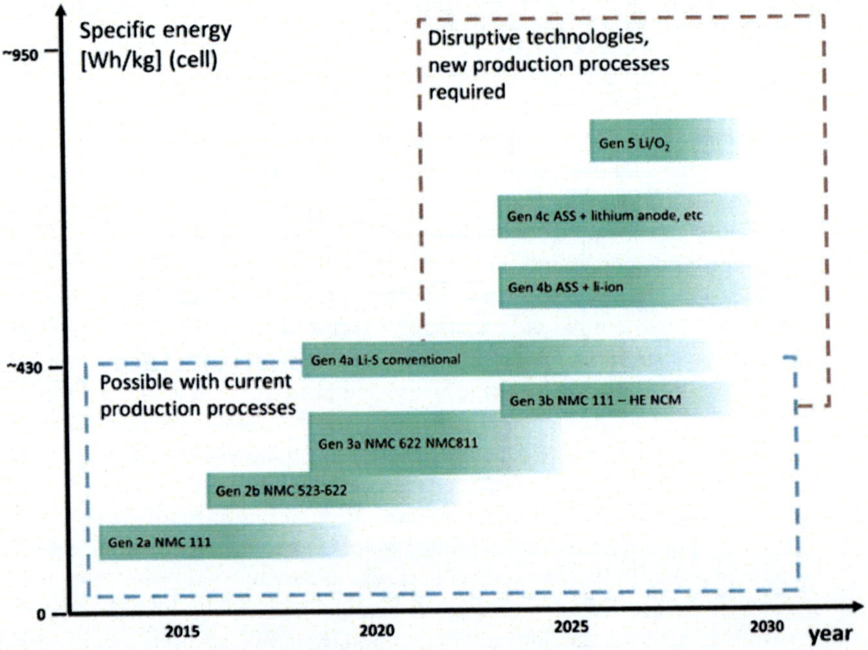

Fig. 1.2 Technology roadmap for traction battery cells. Adapted from (NPE 2016)

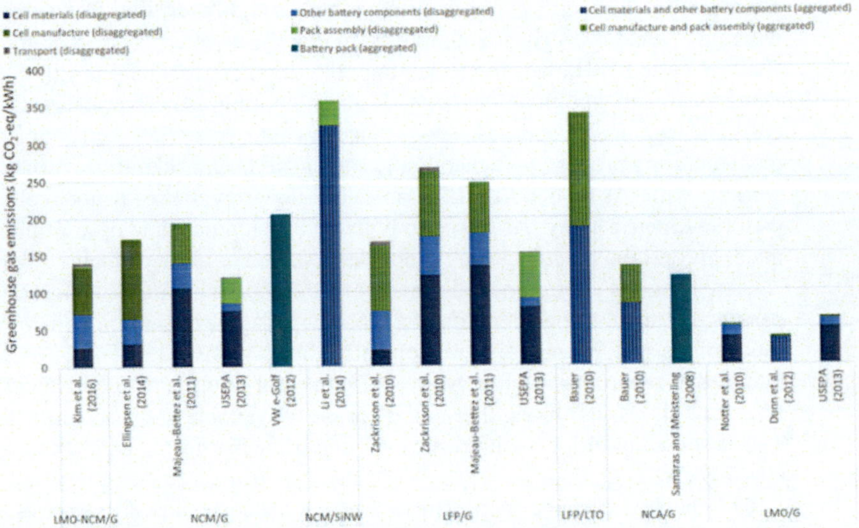

Fig. 1.3 Variation of LCA results for traction batteries throughout the currently available literature (Ellingsen et al. 2017)

a life cycle perspective imply, therefore, an enhanced complexity, as the evaluation itself is very challenging.

To support environmentally based decision making for the further development of electromobility, adequate methods and tools that are able to capture all relevant interactions characterizing the life cycle of an EV are needed. Modelling methodologies dealing with sustainability issues of electric vehicles should focus on capturing the range of possible values a particular impact category can take, rather than aspiring towards obtaining one specific result, while continuing to consider all relevant factors and scenarios in the configuration of the model depicting the product and its life cycle.

1.4 Objective, Context and Structure of This Work

Against the presented background, a new approach for the modelling of the life cycle of complex product systems is introduced. This development is a computational approach that helps to quantify the life cycle environmental impacts of batteries for electric vehicles, while considering a large variability of technical parameters and the effect of related influencing factors. The structure and general outline of the book is shown in Fig. 1.4. Chapter 2 introduces the relevant theoretical background which serves as the foundation for the research.

In Chap. 3, the current state of research is synthetized and analyzed, and the research gap is identified. Chapter 4 introduces the concept of Integrated Computational Life Cycle Engineering of Electric Vehicles and describes a prototypical implementation of the system created. Chapter 5 illustrates the application of the concept with two case studies: one on the analysis of the production of traction batteries; and the second on the application of the concept for the analysis of the use phase of an electric vehicle. In Chap. 6, a critical appraisal is presented and further research directions are summarized.

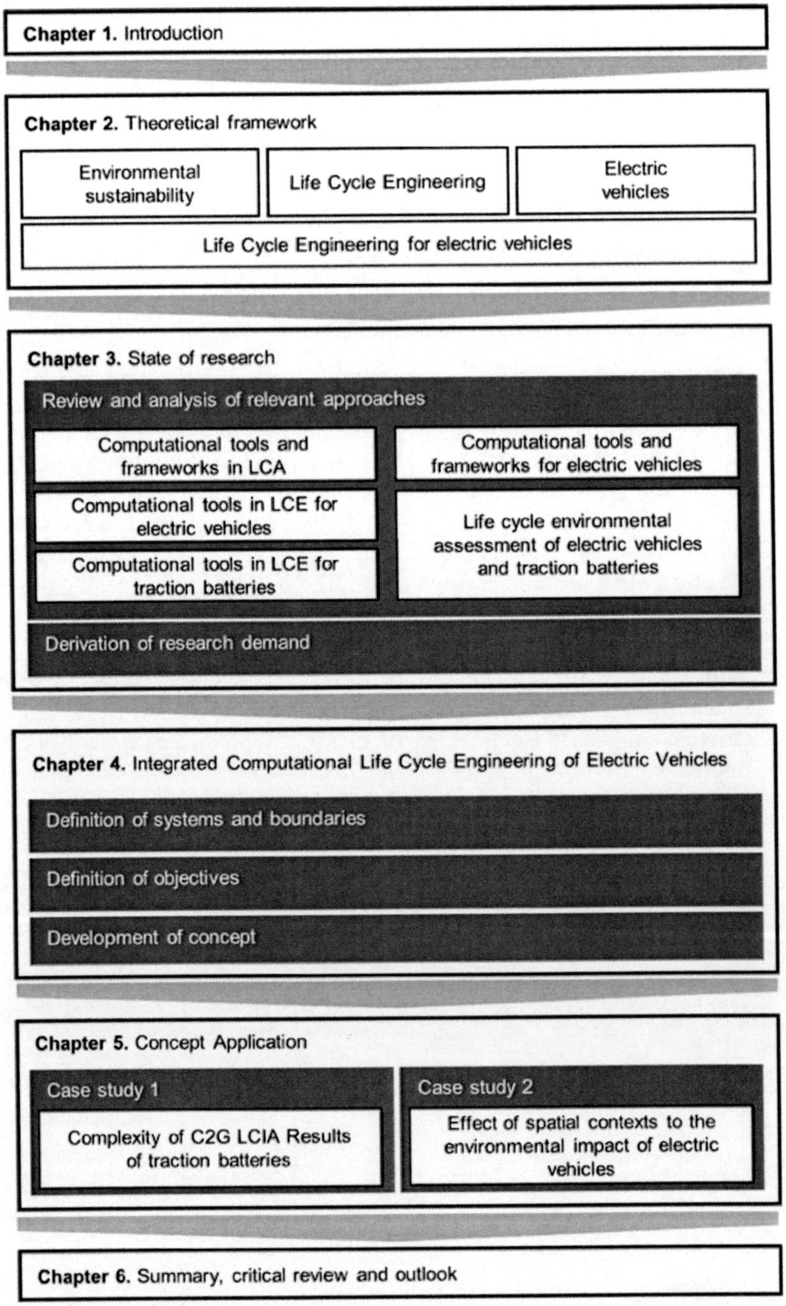

Fig. 1.4 Outline of the book

References

Baitz M, Albrecht S, Brauner E et al (2013) LCA's theory and practice: like ebony and ivory living in perfect harmony? Int J Life Cycle Assess 18:5–13. https://doi.org/10.1007/s11367-012-0476-x

Cerdas F, Andrew S, Thiede S, Herrmann C (2018a) Environmental aspects of the recycling of lithium-ion traction batteries. In: Lithorec, pp 267–288

Cerdas F, Egede P, Herrmann C (2018b) LCA of electromobility. Life Cycle Assess 669–693. https://doi.org/10.1007/978-3-319-56475-3_27

Cerdas F, Titscher P, Bognar N et al (2018c) Exploring the effect of increased energy density on the environmental impacts of traction batteries: a comparison of energy optimized lithium-ion and lithium-sulfur batteries for mobility applications. Energies 11:150. https://doi.org/10.3390/en11010150

Dunn JB, Gaines L, Kelly JC et al (2015) The significance of Li-ion batteries in electric vehicle life-cycle energy and emissions and recycling's role in its reduction. Energy Environ Sci 8:158–168. https://doi.org/10.1039/C4EE03029J

EC-JRC (2010) ILCD Handbook: framework and requirements for life cycle impact assessment models and indicators

Ellingsen LA-WW, Majeau-Bettez G, Singh B et al (2014) Life cycle assessment of a lithium-ion battery vehicle pack. J Ind Ecol 18:113–124. https://doi.org/10.1111/jiec.12072

Ellingsen LAW, Hung CR, Strømman AH (2017) Identifying key assumptions and differences in life cycle assessment studies of lithium-ion traction batteries with focus on greenhouse gas emissions. Transp Res Part D Transp Environ 55:82–90. https://doi.org/10.1016/j.trd.2017.06.028

European Commission (2014) Directive 2006/66/EU on batteries and accumulators and waste batteries and accumulators

Gemechu ED, Sonnemann G, Young SB (2017) Geopolitical-related supply risk assessment as a complement to environmental impact assessment: the case of electric vehicles. Int J Life Cycle Assess 22:31–39. https://doi.org/10.1007/s11367-015-0917-4

Hauschild MZ, Herrmann C, Kara S (2017) An integrated framework for life cycle engineering. Procedia CIRP 61:2–9. https://doi.org/10.1016/j.procir.2016.11.257

Hawkins TR, Singh B, Majeau-Bettez G, Strømman AH (2013) Comparative environmental life cycle assessment of conventional and electric vehicles. J Ind Ecol 17:53–64. https://doi.org/10.1111/j.1530-9290.2012.00532.x

Hellweg S, Mila i Canals L (2014) Emerging approaches, challenges and opportunities in life cycle assessment. Science (80-) 344:1109–1113. https://doi.org/10.1126/science.1248361

Herrmann C (2010) Ganzheitliches life cycle management

International Energy Agency (IEA) (2016) Global EV outlook 2016 beyond one million electric cars

International Energy Agency (2019a) CO_2 emissions from fuel combustion overview. Paris

International Energy Agency (2019b) Global EV outlook 2019. Paris

Kahhat R, Williams E (2012) Materials flow analysis of e-waste: domestic flows and exports of used computers from the United States. Resour Conserv Recycl 67:67–74. https://doi.org/10.1016/j.resconrec.2012.07.008

Karagulian F, Belis CA, Dora CFC et al (2015) Contributions to cities' ambient particulate matter (PM): a systematic review of local source contributions at global level. Atmos Environ 120:475–483. https://doi.org/10.1016/j.atmosenv.2015.08.087

Kwade A, Haselrieder W, Leithoff R et al (2018) Current status and challenges for automotive battery production technologies. Nat Energy 3:290–300. https://doi.org/10.1038/s41560-018-0130-3

McKinsey (2017) Electrifying insights: how automakers can drive electrified vehicle sales and profitability

Michaelis S, Rahimzei E, Kampker A et al (2018) Roadmap Batterie-Produktionsmittel 2030. VDMA

Nelson P, Gallagher KG, Bloom I, Dees DW (2017) Modeling the performance and cost of lithium-ion batteries for electric-drive vehicles chemical sciences and engineering division, second edition. Next Gener Energy Storage Conf 116. https://doi.org/10.3133/fs20143035

NPE (2016) Roadmap for an integrated cell and battery production in Germany 68

Schmidt T, Buchert M, Schebek L (2016) Investigation of the primary production routes of nickel and cobalt products used for Li-ion batteries. Resour Conserv Recycl 112:107–122. https://doi.org/10.1016/j.resconrec.2016.04.017

Schmuch R, Wagner R, Hörpel G et al (2018) Performance and cost of materials for lithium-based rechargeable automotive batteries. Nat Energy 3:267–278. https://doi.org/10.1038/s41560-018-0107-2

Sims R, Schaeffer R, Creutzig F et al (2014) Transport. In: Edenhofer O, Pichs-Madruga R, Sokona Y, Farahani E, Kadner S, Seyboth K, Adler A, Baum I, Brunner S, Eickemeier P, Kriemann B, Savolainen J, Schlömer S, von Stechow C, Zwickle T, Minx JC (eds) Climate change 2014: mitigation of climate change. Contribution of working group III to the fifth assessment report of the intergovernmental panel on climate change. Cambridge University Press, Cambridge, United Kingdom and New York, NY, USA, pp 599–670

Thomas J (2014) Drive cycle powertrain efficiencies and trends derived from EPA vehicle dynamometer results. SAE Int J Passeng Cars Mech Syst 7:1374–1384. https://doi.org/10.4271/2014-01-2562

UNFCCC (2015) Paris declaration on electro-mobility and climate change and call to action. 1

Velten K (2008) Mathematical modeling and simulation. Wiley-VCH Verlag GmbH & Co, KGaA, Weinheim, Germany

Volkswagen AG (2013) The e-mission. Electric mobility and the environment. Production

Wolfram P, Lutsey N (2016) Electric vehicles: literature review of technology costs and carbon emissions. Int Counc Clean Transp 1–23. https://doi.org/10.13140/RG.2.1.2045.3364

Chapter 2
LCE and Electromobility

Contents

This chapter introduces the theoretical background. First, the concept of environmental sustainability and the research field of life cycle engineering are introduced. Next, a general overview on the technical aspects of EVs and traction batteries is presented. This section in particular intends to describe the operating principles of EVs and traction batteries, and to give an overview of their components, materials and manufacturing processes. Finally, the most relevant aspects of electromobility from a life cycle engineering perspective are summarized.

F. Cerdas, *Integrated Computational Life Cycle Engineering for Traction Batteries*, Sustainable Production, Life Cycle Engineering and Management,
https://doi.org/10.1007/978-3-030-82934-6_2

2.1 Environmental Sustainability and Life Cycle Engineering

Environmental sustainability (ES)[1] was defined by Goodland as the "maintenance of the natural capital[2]" (Goodland 1995). Goodland asserted that sustainable development is "development without throughput growth beyond the environmental carrying capacity and which is socially sustainable". This definition, which was originally given by Daly (1990, 1992), has been argued to be less ambiguous than the official definition given by the United Nations (UN),[3] as the latter "does not distinguish among the different concepts of growth and development" (Goodland 1995). While the idea of *environmental protection* is not explicitly mentioned in most of the definitions given for the term *sustainability*, these two concepts (i.e. sustainability and environmental protection) are very often used as synonyms (Moltesen and Bjørn 2018). According to Goodland, sustainability can only be achieved if "the scale of the human economy is kept within the capacity of the overall ecosystem on which it depends". If the natural capital is considered as being infinite, the idea of a sustainable growth (as implied in the official definition of the UN) is therefore self-contradictory (Goodland 1995).

This notion of a society being part of a *finite natural system* that does not grow has given rise to new paradigms of sustainability[4] as the one shown in Fig. 2.1. Under this paradigm, economic development is not seen as a competitor of environmental

---

[1] In this work, the concept of sustainability is dealt exclusively from an environmental perspective. This goes in line with the most contemporary developments within the fields of Sustainable Manufacturing and Life Cycle Engineering. In these fields, efforts are being made towards a transition to an absolute perspective of sustainability within engineering activities (Hauschild 2015; Hauschild et al. 2017a; Kara et al. 2018).

[2] Natural capital is defined as a stock of natural assets (i.e. natural resources such as soil, atmosphere, forests, etc.) which provide useful goods and/or services (Goodland 1995; Goodland and Bank 2002).

[3] Sustainable development was officially defined in the so-called Brundtland report (World Commission on Environment and Development 1987) as development that "meets the needs of the present without compromising the ability of future generations to meet their own needs". The three components of sustainable development (social development, economic development and environmental protection) were further recognized and integrated into the definition as being "interdependent and mutually reinforcing" (United Nations 2002). This understanding of sustainable development has been described as weak (Parliamentary Comissioner for the Environment 2002; Keiner 2006; IUCN 2008) because it encourages a competition between the three dimensions. This has been argued to allow the existence of trade-offs between them which implies, for instance, that environmental impacts might be justified or compensated by social and/or economic improvements (Pearce and Atkinson 1993, 2010; Ayres et al. 2001; Gutowski 2011).

[4] A stronger vision of sustainability, as presented by Goodland and Pearce (Pearce and Atkinson 1993; Goodland 1995) and Ayres and colleagues (Ayres et al. 2001), establishes that although natural capital is an essential input to economic production and social welfare, it cannot be substituted by physical or human capital. In this line of thought, economy is seen as a subsystem of society, meaning that economy only exists if society exists and society can only exist if the environment exists (Rockström 2015).

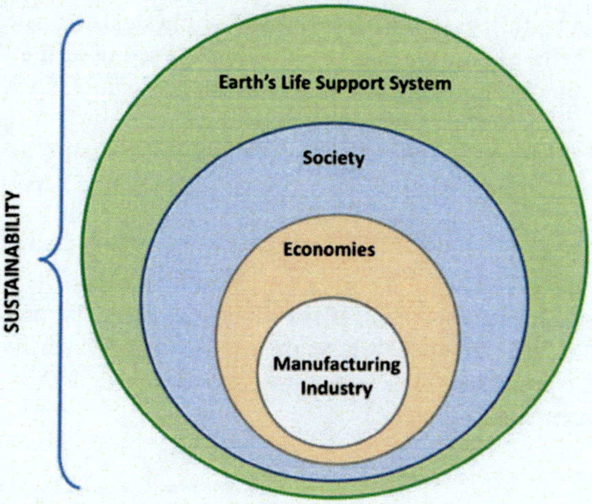

Fig. 2.1 Sustainability and its dimensions within the paradigm of a sustainable development for the Anthropocene.[5] From Hauschild (2017b), based on Rockström (2015)

protection but rather as a way to achieve social welfare within the carrying capacity of the earth (Rockström 2015).

Within this paradigm, although the term ES makes direct reference to the environment, its main motivation is the protection of human life (Goodland 1995). ES is therefore an anthropocentric perspective of sustainability aiming at conserving all earth's life-support systems indefinitely. This implies reducing or avoiding the environmental impacts rising from human activities.

The following sections defines environmental impacts and their relationship to human activity (2.1.1), how to estimate them (2.1.2) and the role of engineering towards reducing or avoiding them (2.1.3).

2.1.1 *Environmental Impacts of Products*

An environmental impact is generally defined as an impact on natural systems, human health systems, or the depletion of resources caused by an interaction between a system in the technosphere (i.e. everything that is manmade) and the ecosphere (i.e. everything that is not manmade, also referred to as "environment" or "nature") (European Commission—Joint Research Centre—Institute for Environment and Sustainability 2010). These exchanges[6] can be understood as flows of material and energy passing through the boundary between the manmade system and nature (ISO 2006).

[5] Geological epoch in which humankind has surpassed in relevance other geological forces shaping the biosphere (Crutzen 2002; Rockström 2015).

[6] Also referred to as elementary flows, see Fig. 2.2.

They can be further defined as chemical or physical exchanges, such as substances emitted to the air into water bodies or left in the soil or on the soil surface; substances that are removed from the soil; physical changes to flora, fauna and soil; and physical changes to the surface (Weidema 2001). Through a series of physical, chemical and biological processes,[7] these exchanges lead to set final effects such as damages to the human health, to the ecosystem quality, loss of biodiversity, loss of resources and damages to manmade environments (de Haes et al. 1999).

While the type of the impact (I) is defined by the type of elementary flows generated by the provision of a particular unit of human activity, its magnitude is driven by the size of the population (P) to which that particular product or service is made available, their affluence (A) and the environmental intensity of the technology (T) applied to provide it. This dependency was expressed in form of an identity equation by Holdren and Ehrlich as:

$$I = \mathrm{P} \cdot \mathrm{A} \cdot \mathrm{T} \tag{2.1}$$

in which is a measure of the per capita impact of a particular technology (Holdren and Ehrlich 1971). This implies that, under the condition of a constant level of consumption, the environmental impact per capita of a particular human activity depends on the environmental impact generated by the technical (product) system applied to provide it[8] (Holdren and Ehrlich 1971; Hauschild 2015; Moltesen and Bjørn 2018).

A technical system is linked to environmental impacts caused by the exchange of elementary flows in each of its *life cycle* stages, meaning from the extraction of the raw materials required through manufacturing, transport and usage to its final disposal treatment (Alting and Jøgensen 1993; Bhander et al. 2003; Rebitzer et al. 2004; Herrmann 2010; Hellweg and Mila i Canals 2014; Hauschild 2018). These exchanges are additionally influenced by local or regional natural conditions (e.g. weather, altitude, etc.) and by the technological landscape in which the technical system operates (e.g. electricity mix, recycling infrastructure, transportation infrastructure, etc.). Furthermore, the increase of production volumes driven by a rise of affluence or population growth also has a significant effect on the total impact of a technical system, and therefore these factors shall be also considered (Kim and Kara 2014; Hauschild et al. 2017a; Kim et al. 2017).

Accordingly, life cycle thinking is key to consistently estimate the total impact of a given technology, but also to develop solutions that help reduce its impact while

---

[7] Also referred to as environmental mechanism. A chain of environmental mechanisms defines the impact pathway (cause-effect chain) of a group of emissions.

[8] A very thorough presentation of the IPAT identity is given by Chertow (2000). In her contribution she gives a historical perspective of the identity and describes the many different interpretations that different groups of environmentalists have made. The key message in her paper is that technology "although associated with both disease and cure for environmental harms, is a critical factor in environmental improvement", and thus it is important and necessary to maintain the efforts on finding the "cure".

avoiding the risk of shifting the impact to another life cycle stage, geographical region, type of impact or even to another space in time.

2.1.2 *Life Cycle Assessment*

Life Cycle Assessment (LCA) is a system analysis and assessment methodology that enables the quantification of potential environmental impacts of product systems by considering all of its life cycle stages (ISO 2006; European Commission—Joint Research Centre—Institute for Environment and Sustainability 2010; Klöpffer and Grahl 2014; Hauschild 2018). As summarized by Baitz and colleagues, the methodology has been internationally acknowledged in industry as being the "best available methodology to investigate environmental sustainability performance in a reliable and transparent way" (Baitz et al. 2013).

A product system[9] (Fig. 2.2) is defined by ISO as a collection of processes performing one or more functions and modelling the life cycle of a product (ISO 2006). The smallest elements of a product system for which data as inputs and outputs is available is called unit process (Klöpffer and Grahl 2014). In practice, a unit process can have different scales and may therefore represent a particular process (e.g. one specific manufacturing process in a factory) or even an whole factory (Bjørn et al. 2018a).

While a product system might be composed of a very large number of unit processes, the scope of a LCA study is normally focused on one (or a set of) particular process for which primary data is available. These processes are usually referred to as being specific to the product system (Hauschild et al. 2017b; Bjørn et al. 2018a) and the system they form is called a *foreground system* (Klöpffer and Grahl 2014; Lesage and Muller 2017). The *background system,* in turn, is composed by all the unit processes to which the unit processes in the foreground system are linked. These unit processes are considered to be of "general concern" (e.g. commodities, electricity, transportation, etc.) and are most of the time modelled using generic data from commercial databases (Klöpffer and Grahl 2014; Hauschild et al. 2017b; Bjørn et al. 2018a).

Notice that, as represented in Fig. 2.2, unit processes belonging to both the foreground and the background system are spatially located, i.e. the spatial context (SC) in which a unit process exist, connects both the background and the foreground systems. This implies not only that the magnitude of the exchanges between the technosphere and the ecosphere are influenced by the specific local or regional contexts (both the natural and the man-made contexts) but also the environmental mechanisms converting these exchanges into impacts (in the case of non-global impacts) may vary spatially (Verones et al. 2015).

---

[9] Also referred to as "*system*" since in LCA different cases can be studied that are not necessarily related to a specific product (for example the fulfillment of needs like mobility) (European Commission—Joint Research Centre—Institute for Environment and Sustainability 2010).

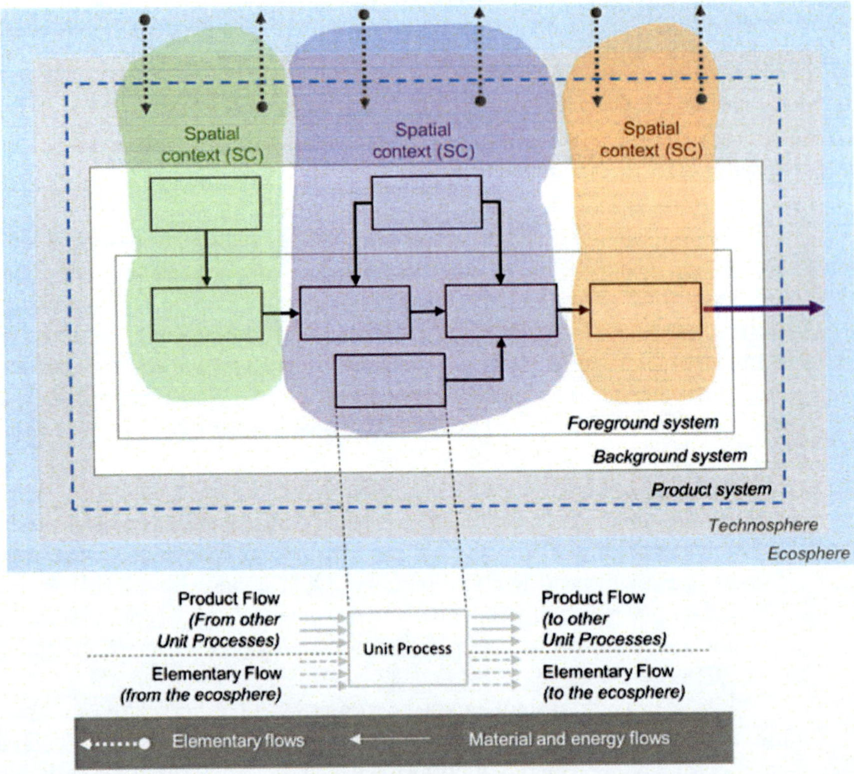

Fig. 2.2 Product system, foreground system, background system and unit process. Own figure, inspired on (EC-JRC 2010a; Klöpffer and Grahl 2014; Bjørn et al. 2018a)

LCA consist of four different phases: (i) goal and scope definition, (ii) life cycle inventory (LCI), (iii) life cycle impact assessment (LCIA) and (iv) interpretation (ISO 2006). These four phases are completed through iterative loops in each of the phases. These iterations aim at calibrating all modelling choices made while filling data gaps or performing consistency checks (EC-JRC 2010a).

During the first phase, *goal and scope definition,* the case study is set up. Here, the functional unit (the unit of function provided by the product system) is selected and the boundaries (geographical, temporal and technological) of the product system are defined (Klöpffer and Grahl 2014; Hauschild et al. 2017b; Bjørn et al. 2018a). The unit processes identified are then classified into foreground and background system and finally the type of environmental impacts intended to be investigated are defined (Hauschild et al. 2017b; Bjørn et al. 2018a).

The *LCI* phase aims at collecting all data required for the study and building the so-called life cycle model. Data collection is understood here as the collection of primary data (i.e. by means of measurements, simulations, experimental models) and secondary data (i.e. data from reports, interviews and scientific papers) for the foreground system and as the selection and identification of generic data for the background system (i.e. searching for datasets,or a proxy of it, within the commercial LCI data available). Once data has been collected, a life cycle model is created. Such a model is thought of as a representation of the complete life cycle stages of the functional unit provided by the product system (Lesage and Muller 2017). It has the objective of estimating the total output of elementary flows across all unit processes linked within a product system (Lesage and Muller 2017). The two most common approaches followed in this regard are the sequential and the matrix approach (Heijungs and Suh 2002).[10]

The third phase is *LCIA*. The objective of this phase is to translate the inventory obtained in the LCI phase into an environmental profile that represents the impact of the product system within specific impact categories (Hauschild and Huijbregts 2015). An impact category is defined by ISO as a "class representing environmental issues of concern" (ISO 2006). The LCIA phase is divided into five steps: (a) selection of impact categories, category indicators and characterization models, (b) classification of LCI results into the selected impact categories, (c) estimation of category indicator result[11] (characterization), (d) normalization and finally (e) weighting, from which only a, b and c are mandatory (Hauschild and Huijbregts 2015). The indicator selected to represent the impact category chosen can be, according to ISO, placed anywhere in the impact pathway between the elementary flow and the areas of protection (AoP) as shown in Fig. 2.3 (ISO 2006). AoPs are technically defined as "clusters of category endpoints of recognizable value to society" (Hauschild and Huijbregts 2015). A midpoint indicator is located somewhere in the middle of the impact pathway, whereas the endpoint indicators (sometimes also called damage oriented indicators) are placed closed to the areas of protections (EC-JRC 2010b; Hauschild and Huijbregts 2015).

The fourth phase of the LCA methodology is called *interpretation* (ISO 2006; Klöpffer and Grahl 2014; Hauschild et al. 2018). The objective in this phase is to perform an analysis of the results considering all of the important aspects and modelling choices of the preceding phases. This phase is divided into three steps.

[10] In the sequential approach each of the unit processes are scaled sequentially (i.e. one at a time) in order for them to deliver the amount of product that the next upstream process requires. (Heijungs and Suh 2002; Lesage and Muller 2017). While this approach is considered to be easier to understand, it is difficult to present a formal mathematical demonstration for product systems with high number of unit processes. Additionally, the approach is normally criticized due to its limitations when dealing with circular or reciprocal dependencies (e.g. electricity is required to produce aluminum and aluminum is required to produce electricity) (Lesage and Muller 2017).

In the matrix approach each of the flows (i.e. elementary flows and flows between unit processes) are mathematically expressed as a system of equations for which output is the functional unit. The advantage of the matrix approach is that by inverting the matrix the circular dependencies are reflected in the final output. A detailed explanation of the mathematical structure of LCI modelling is presented by Heijungs and Suh in (Heijungs and Suh 2002).

[11] In this step a characterization factor (CF) is multiplied to the amount of an elementary flow assigned to a particular impact category. A CF is a value that represents the significance of a specific elementary flow to an impact and enables expressing the contribution of several different elementary flows using a common metric (e.g. kg of CO_2-eq) (Hauschild and Huijbregts 2015).

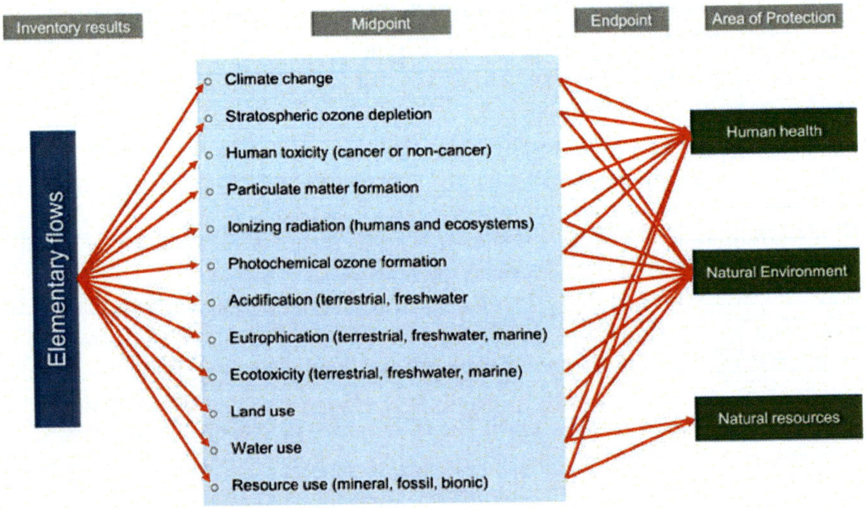

Fig. 2.3 Framework of impact categories for characterization modelling at midpoint and endpoint levels. Adapted from the ILCD guidelines (EC-JRC 2010b)

First, all significant issues (important elementary flows, possible trade-offs, key processes, etc.) are identified and analyzed. Second, there issues are evaluated with regard to their consistency and completeness. An important part of this step is the evaluation of uncertainties.[12]

In summary, LCA is a powerful methodology for the assessment of environmental impacts of product systems (Bjørn et al. 2018b) and for the development of less environmentally intensive technologies (Bhander et al. 2003; Bonou et al. 2015).

2.1.3 *Life Cycle Engineering*

Definitions and scope

Life Cycle Engineering (LCE) has its roots in concerns from a production engineering perspective regarding environmental issues such as waste generation and damages to the natural resources (Herrmann et al. 2014). Since its origins, LCE

[12] A comprehensive overview of the importance of considering uncertainties in LCA, their sources and types, is presented in (Rosenbaum et al. 2018). For information regarding the available (common) methodologies to examine error propagation and uncertainties in LCA, the reader may refer to the work presented in (Huijbregts et al. 2003; Groen et al. 2014; Heijungs and Lenzen 2014; Gregory et al. 2016; Beltran et al. 2018). A methodology to estimate uncertainties in LCA with pre-calculated aggregated datasets is presented in (Lesage et al. 2018). Information on methods to perform sensitivity analysis in LCA are presented in (Wei et al. 2015). More recent work on the integration of global sensitivity analysis in LCA including some case studies can be found in (Groen et al. 2017; Lacirignola et al. 2017; Sacchi et al. 2019).

has been given several definitions. Alting and Legarth defined LCE as "the art of designing the product life cycle through choices about product concept, structure, materials and processes" (Alting and Legarth 1995). Westkämper and colleagues (Westkämper et al. 2000) brought the concept of sustainability into the definition by arguing that the purpose of LCE is to contribute towards "compliance with the key issues of sustainable development". Early definitions of LCE were merely focused on environmental efficiency (or eco-efficiency). Its purpose was described as aiming to minimize the environmental impact of a product during its entire life cycle (Wanyama et al. 2003) and at "protecting the environment and conserving resources, while encouraging economic progress, keeping in mind the need for sustainability, and at the same time optimizing the product life cycle and minimizing pollution and waste" (Jeswiet and Hauschild 2005). The sense of efficiency expressed in the definitions above was stressed later by Hauschild and colleagues (2005) when they stated that the aim of LCE is to "improve the eco-efficiency of industrial activities". Eco-efficiency[13] is further defined in their paper as the ratio between a service provided by an industrial activity and its environmental impact (Hauschild et al. 2005). In an effort to re-focus the scope of LCE towards an absolute perspective of sustainability (see Fig. 2.1), Hauschild and colleagues presented an integrated framework for LCE [subsequently named as the Lyngby Framework (Kara et al. 2018)], in which the environmental dimension is seen as an independent finite system which provides the foundation for social and economic sustainability (Hauschild et al. 2017a). As represented in Fig. 2.4, their framework aims to bring together both top-down and bottom-up approaches in LCE so as to provide support towards eco-effective technical solutions. In this framework, the environmental concern includes the scale of environmental impacts of product systems considering its temporal and spatial scale.

Along with the Lyngby Framework, a new definition for LCE was given. LCE is now defined as "sustainability-oriented product development activities," and the methods and tools supporting LCE are used to support reduction of the environmental impact of technologies in an absolute manner, considering increases in affluence, population and the Earth as a finite system (Bjørn et al. 2015; Kara et al. 2018).

Methods and Tools for LCE

Contrary to the understanding of sustainability under which LCE has been defined, its general field of action has remained to a large extent unchanged. In an early contribution on LCE, Ishii refers to the field as the decisions made during manufacturing and design that have a significant impact on the product's life cycle (Ishii 1995). While Ishii's perspective was merely cost-driven, he recognized the identification of life cycle knowledge to perform design evaluations as a major task and challenge in LCE (Ishii 1995). Similarly, Kara roughly describes the task of LCE as "applying life cycle knowledge to engineering solutions" (Kara 2009). *Life cycle*

[13] A detailed discussion on eco-efficiency is presented in (Hauschild 2015). In their contribution the authors discuss why focusing exclusively on increasing the eco-efficiency of industrial activities is not sufficient to reduce the total environmental impact of products.

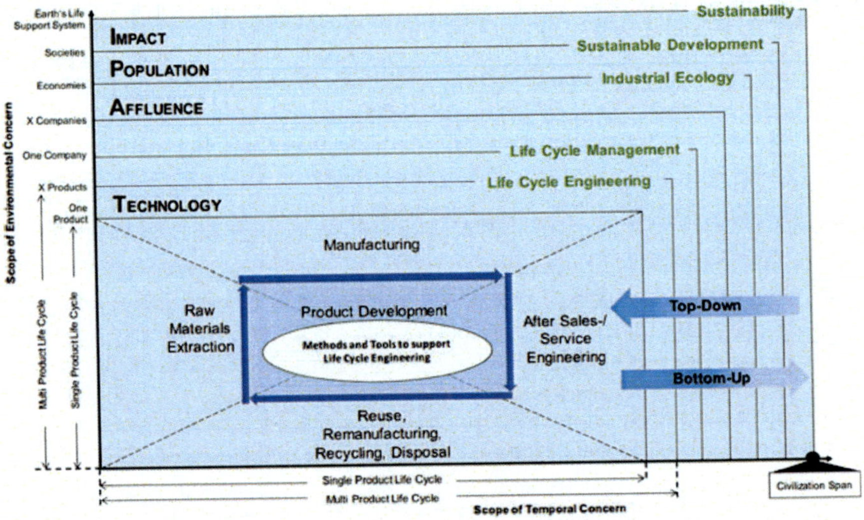

Fig. 2.4 LCE framework, also known Lyngby Framework (Hauschild et al. 2017a)

knowledge has been defined[14] as the specific knowledge and experience found in each product life cycle stage that is relevant for product design (Herrmann 2010). The LCE research community describes this life cycle-oriented exchange of information as the integration of environmental parameters within product design, development and manufacturing (Alting and Legarth 1995; Westkämper et al. 2000; Jeswiet and Hauschild 2005) for the sake of discovering improvement potentials (Westkamper et al. 2000) and being able to draft measurable targets and technical solutions (Broch et al. 2015). A representation of the life cycle-oriented information and knowledge flow for product development can be referenced in Fig. 2.5.

Mansour (2006) categorized this knowledge into: (i) requirements (e.g. environmental regulations and standards, recycling rates, etc.), (ii) feedback regarding advantages and disadvantages of the technical solutions or design (from each life cycle stage, e.g. recyclability, information regarding the environmental impact of specific materials, requirements from the usage stage, relevant process parameters, etc.) and (iii) experience or overall performance from the complete life cycle. An important number of methods and tools have been developed to make use of this information in

[14] In his book, Herrmann uses the term "Life Cycle Design relevant knowledge" (Herrmann 2010).

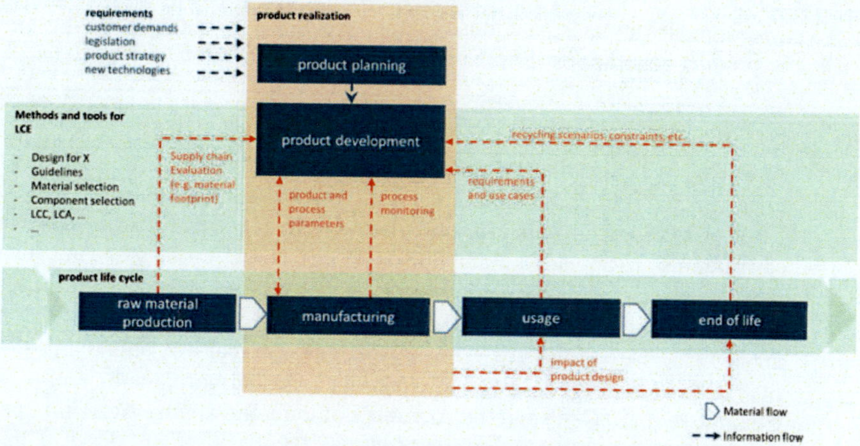

Fig. 2.5 Life cycle-oriented information and knowledge flow for product development. Based on (Abele et al. 2005; Mansour 2006; Herrmann 2010)

decision making for LCE. Hauschild and colleagues (2017a) classified[15] the methods and tools available into two groups:

- Specific tools developed to support one particular life cycle stage with influence on the whole product life cycle (i.e. concurrent (eco-)design tools and methodologies). These tools range from guidelines, checklists, diagram tools and metrics, to integrated approaches and modelling software such as CAD integrated tools (Westkamper et al. 2000; Rossi et al. 2016). Some examples of methodologies in this group are specific guidelines for component or material selection and design methodologies including their software support[16] such as Design for Environment (DfE), Design for Recycling (DfR) and Design for Disassembly (DfD), among others.
- Generic tools that can be used at any life cycle stage such as LCA and Life Cycle Costing (LCC). Due to its complexity and its time and resource intensity (Keoleian 1993), LCA, rather than being applied as a concurrent design tool, is seen as a retrospective tool (Telenko et al. 2008). This implies that LCA has been traditionally applied in the last phases of the development process after a detailed

[15] Other classifications of methods and tools for LCE have been presented by Duflou and colleagues (2003), Abele and colleagues (2005), and Ishii (1995) in terms of the type of feedback (anticipative and influential), the applicable stage in the product life cycle and the applicable stage in the design or development process; Bovea (Bovea and Pérez-Belis 2012) presents a classification of methods and tools based on 5 criteria: (1) method applied for the environmental assessment study, (2) the product requirements, (3) scope (one life cycle stage or the whole life cycle), (4) type of results offered (i.e. quantitative or qualitative), (5) applicable product design stage.

[16] See, for example, the work done by Ashby and colleagues in (Ashby 2013; Granta Design Limited 2013) in the field of environmentally-informed materials selection, and Herrmann (Herrmann 2003) in the field of DfR.

design already exist and mostly with the aim of providing feedback towards next product generations (Keoleian 1993; Ishii 1995; Duflou et al. 2003).

LCA-based methods and tools for LCE

As argued by Hauschild and colleagues (2017a), technological products are required to be life cycle engineered. In this regard, the methods and tools for LCE need to be based on LCA as shown in Fig. 2.5.[17] LCA has been at the core of LCE ever since the early definitions appeared. For Alting and Legarth, LCA is a unique holistic methodology used to estimate the environmental consequences of these engineering choices (Alting and Legarth 1995). Westkämper and colleagues posit that (Westkämper et al. 2000) LCA methodology is especially valuable in that it is able to "provide quantitative basic data for a sustainable product". Furthermore, the Scientific Technical Committee *A* (STC-A) and the Assembly of the International Academy for Production Engineering (CIRP as referred to by its French acronym) describe the field of LCE as research that "covers Life Cycle Assessment of products, manufacturing processes and systems" (CIRP 2017).

Several characteristics have contributed to make LCA the most appropriate instrument to investigate the environmental impacts of product systems. First, the methodology enables a systematic modelling of a product's life cycle. Taking a life cycle perspective allows for the identification of possible problem shifting among processes, life cycle stages or impact categories (Hellweg and Mila i Canals 2014; Bjørn et al. 2018b). Second, LCA provides a basis for assessing product systems quantitatively (Bjørn et al. 2018b), generating the results required to make numerical comparisons or rankings and to be integrated into other engineering models (e.g. parameter optimizations, process or product design, material selection among many others). Finally, LCA is based on scientific principles (Bjørn et al. 2018b). All characterizations models in the LCIA phase that describe the relationship between emissions and damages are based on demonstrated causalities or on empirical observations (Bjørn et al. 2018b). Moreover, the data gathering and modelling in the LCI phase is based on solid laws of nature (e.g. conservation of mass, first and second law of thermodynamics and principles of stoichiometry) (Klöpfer and Grahl 2014) and the life cycle model as well as the evaluation of uncertainties are based on robust economic and mathematic methods (Heijungs and Suh 2002).

---

[17] This thesis builds upon the LCA-based perspective of LCE presented in (Hauschild et al. 2017a) and therefore it does not go into much details to describe the specific (eco-) design-oriented methodologies. To this end the reader may refer to the contribution from Telenko and colleagues (Telenko et al. 2008) for a comprehensive overview on DfE and DfR methodologies. An overview of over 600 available tools for eco-design classified according the type of feedback and implementation purpose is given in (Rousseaux et al. 2017). A meticulous review and classification of eco-design tools and methods is presented by Rossi and colleagues (Rossi et al. 2016). In Lindahl and colleagues (Lindahl and Ekermann 2013), a structure to categorize these tools and methods is provided in order to benchmark them in terms users' needs and potential applications. For information on LCC, the reader might refer to the works from Woodward and Dhillon (Woodward 1997; Dhillon 2009).

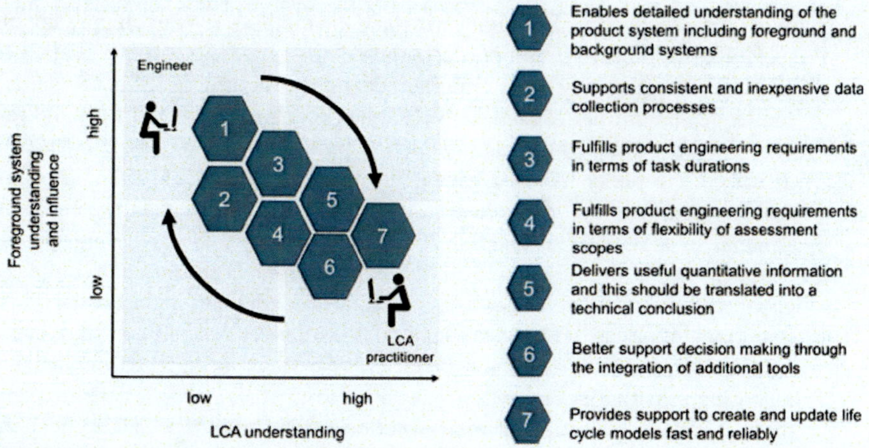

Fig. 2.6 Challenges of LCA in engineering regarding the degree of expert knowledge required (own figure). This list is not intended to be exhaustive

There are nevertheless several obstacles still preventing LCA to become a mainstream instrument in engineering and decision making[18] as summarized in Fig. 2.6 (Bhander et al. 2003; Baitz et al. 2013; Curran 2014; Laurin et al. 2016). These can be grouped into methodological, applicability and interpretability challenges. The *methodological issues* are mostly related to the complexity of both the product system and the methodology itself. Ideally, in order to perform a LCA, a comprehensive understanding of both the product system's technical context and the LCA methodology is required. The engineering reality of LCA, however, is often characterized instead by a division of capabilities concerning the degree and diversity of expert knowledge (Bey et al. 2013), which in most cases demands an intensive interdisciplinary team.

From the perspective of the decision maker, LCAs can be in general very complex and require intensive assistance from environmental or LCA experts (Bhander et al. 2003; Cha and Suh 2011; Cerdas et al. 2017). Particular challenges in this regard are the selection of the different impact categories, including their cause-effect chain models (Keoleian 1993; Cerdas et al. 2017); ensuring consistency while defining system boundaries and functional units (Bhander et al. 2003); and the understanding of the data requirements and availability for the background system, as well as their modelling approaches, representativeness and uncertainty. From the perspective of an LCA practitioner, the most important obstacle is usually an insufficient understanding of the foreground system which, if there is deficient communication or limited access to primary and/or secondary data (Keoleian 1993), ends up reflected in the quality and reliability of the results. Finally, LCA practitioners rarely understand the science

[18] While the range of application of LCA varies from providing support during product and process development activities to marketing and policy making (Owsianiak et al. 2018), decisionmaking in the context of this thesis is addressed exclusively from an engineering perspective.

of decision making within engineering and the methodology itself does not offer any systematic approach to support decision making processes (Laurin et al. 2016).

LCA in engineering faces important challenges regarding its *applicability*. These challenges are mostly related to the LCI phase due to matters of time and resources intensity in regards to the data collection processes and its detachment from time to market (Keoleian 1993; Bhander et al. 2003; Bey et al. 2013; Curran 2014), but also due to limitations when it comes to updating and maintaining the models (Bhander et al. 2003; Cha and Suh 2011). Additionally, LCI modelling approaches in engineering are often criticized for delivering a limited snapshot of a complex system and therefore are not able to consider all interdependencies in the design of a product development process (Bhander et al. 2003). Moreover, simplified and sector specific LCA tools are often created to provide an assessment of the environmental impacts regarding a particular research question[19] (Rossi et al. 2016). These simplifications often involve a reduction of the resolution of the input data, simplification of the calculation models, friendly (high-level) user interfaces and a high aggregation and/or weighting of results (Rossi et al. 2016). Aggregation andsimplifications are often criticized for distorting (Keoleian 1993) and lending a high degree of uncertainty to the presentation of results (Lesage et al. 2018).

LCA faces several issues regarding the *interpretability* of the results. The methodology does not support quick, reliable identification of important components or processes within the product system (Bhander et al. 2003) and, what's more, the results (i.e. environmental consequences) are seldom connected to a root cause (i.e. as a technical conclusion) (Baitz et al. 2013). In addition, the data used and the modelling choices done are often highly uncertain and far from representing the industrial reality (Baitz et al. 2013). Finally, LCA is not always able to indicate a clear advantage when comparing different alternatives as the methodology does not provide a guide to analyze trade-offs systematically (Laurin et al. 2016). In short, a seamless implementation of LCA within engineering activities depends largely on an efficient *computational support,* able to handle all relevant research questions and to integrate interdisciplinary knowledge in order to capture and interpret the high complexity of the interactions between technology and the environment.

2.2 Technical Aspects of Electric Vehicles and Lithium-Ion Traction Batteries

2.2.1 Electric Vehicles: Definitions and Classification

The term electric vehicle (EV) is generally defined as a vehicle that is partly or completely propelled by an electric motor and that relies on an electrochemical

[19] See for example the models developed by (Nordelöf and Tillman 2018) to create life cycle inventories and assessments of electric motors.

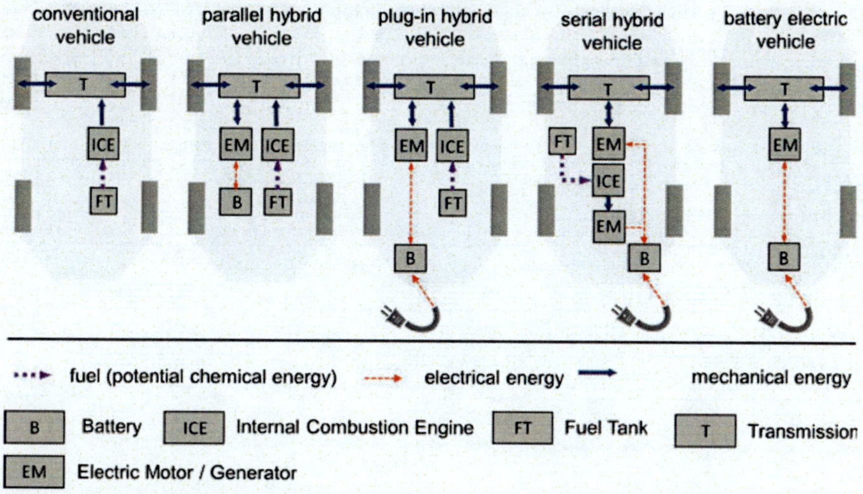

Fig. 2.7 Different configuration of vehicle propulsion systems. Own Figure based on (Cerdas et al. 2018a, 2018b, 2018c, 2018d)

or electrostatic system to store all or part of the energy demanded (Guzzella and Sciarretta 2013). Figure 2.7 shows a general description of the setups of different types of propulsion systems for vehicles, and the different types of energy flows between the components.

EVs can be classified into hybrid electric vehicles (HEV) and battery electric vehicles (BEV), depending on the degree of hybridization of the power train (Guzzella and Sciarretta 2013). HEVs use a combination of an engine (ICE) and an electric motor (EM) for its operation (Guzzella and Sciarretta 2013). These are usually characterized following two different criteria. Regarding the configuration of the propulsion system, HEVs are classified into serial and parallel HEV (Fig. 2.7). In the serial HEV, the ICE is coupled to a generator that converts the mechanical energy into electrical energy, which is used to charge a battery system (B). The traction energy is then delivered to the wheels entirely by one or several EMs, which are fed with the electrical energy from the battery system (Hofmann 2010). Parallel HEVs are characterized by having a direct mechanical connection between the transmission and both the EM and the ICE. In this way, parallel HEVs are capable of operating entirely propelled by the electric system, the internal combustion system or a combination of both propulsion systems (Hofmann 2010). The degree of hybridization (DoH), is defined as:

$$DoH = \frac{P_{b,\max}}{P_{b,\max} + P_{e,\max}} \tag{2.2}$$

and being $P_{b,\max}$ and $P_{e,\max}$ the maximum power capable to be delivered by the battery and the ICE respectively (Guzzella and Sciarretta 2013). According to the

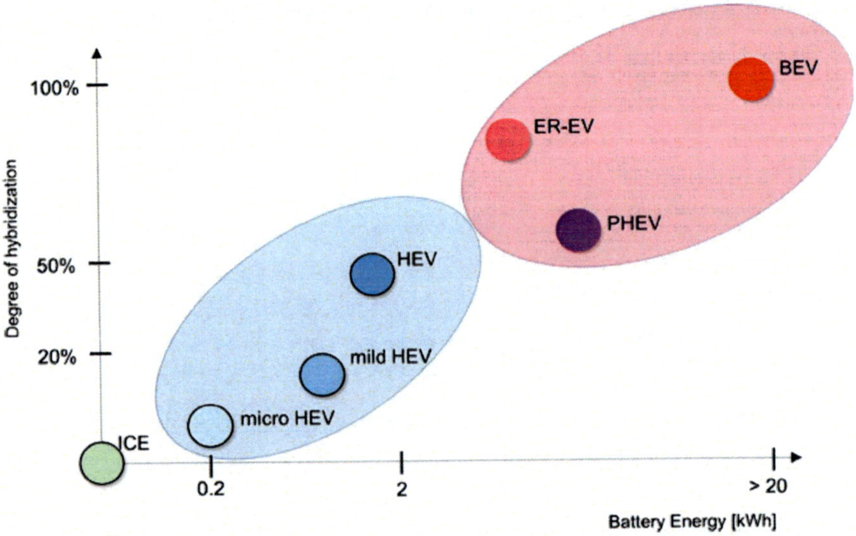

Fig. 2.8 Typical values of the DoH for EV. Based on (Guzzella and Sciarretta 2013). HEV: Hybrid Electric Vehicle, PHEV: Plug-in Hybrid Electric Vehicle, ER-EV: Extended-Range Electric Vehicle, BEV: Battery Electric Vehicle

DoH, EVs can be divided into micro hybrid, mild hybrid, full hybrid, plug-in hybrid electric vehicles (PHEV) and extended range electric vehicles (EREV) (Guzzella and Sciarretta 2013; Tschöke 2015). Typical DoH values are shown in Fig. 2.8.

Although there are no clear limits to classify HEVs based on the DoH, a general characterization is done according to the level of vehicle functionality possible through electric energy (Fig. 2.9). A higher DoH implies that the benefits of EVs are evident, regarding reduction of energy consumption in the usage stage and a reduction of tailpipe emissions per kilometer. This also means, however, that draw-backs such as manufacturing cost and a potential perceived loss of driving autonomy increase (Guzzella and Sciarretta 2013).

Micro HEVs have the lowest degree of hybridization. These types of vehicles are basically conventional engine vehicles equipped with an extra electric motor connected in parallel to support automatic start-stop routines of the main engine (Guzzella and Sciarretta 2013). The electric energy is provided by a lead acid battery or a standard 14 V generator (Tschöke 2015). The electric power train of the mild HEVs allows limited energy recuperation and engine boosting, however full electric driving is not possible. Full HEVs are able to power all their operations modes with the electric power train. These systems are further classified into autarchic full HEV (usually called full hybrid or HEV) and plug-in hybrid electric vehicles (PHEV) (Tschöke 2015). The battery system in PHEVs can be recharged with electricity from the grid. In these vehicles, the DoH is usually higher than the autarchic HEV, meaning that the size of the battery system, and therefore its charge-depleting operation range, is larger [usually ranging from 30 to 100 km (Hofmann 2010; Tschöke 2015)].

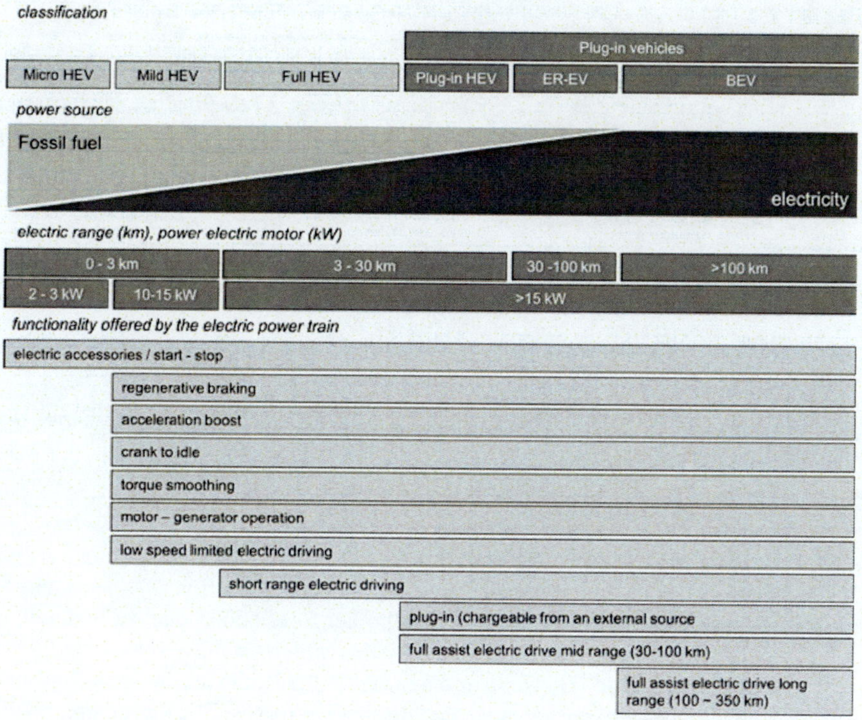

Fig. 2.9 EVs classification according to the functionality of the electric power train. Based on (Karden et al. 2007; Hofmann 2010; Guzzella and Sciarretta 2013; Herrmann and Rothfuss 2015)

Another type of PHEVs is the so called extended-range electric vehicles ER-EV. These are serial HEVs that can be recharged with both power from the grid and also with an on-board auxiliary power unit (APU) that enables the vehicle to increase the electrical range (Guzzella and Sciarretta 2013).

Finally, BEVs are solely powered by electricity and are composed of an electric motor, an energy storage system and a set of electronic components controlling the electricity flow between the vehicle elements (Guzzella and Sciarretta 2013).

2.2.2 *Electric Vehicle Energy Demand*

Disregarding the configuration of the power train, EVs are required to overcome a set of forces that act on the vehicle depending of its geometric characteristics (e.g. mass) and driving conditions. These are represented in Fig. 2.10. As shown, the energy consumed at any given moment by a vehicle while driving depends on the friction losses caused by the aerodynamic and rolling friction forces, in addition to the energy dissipated by the brakes (Guzzella and Sciarretta 2013).

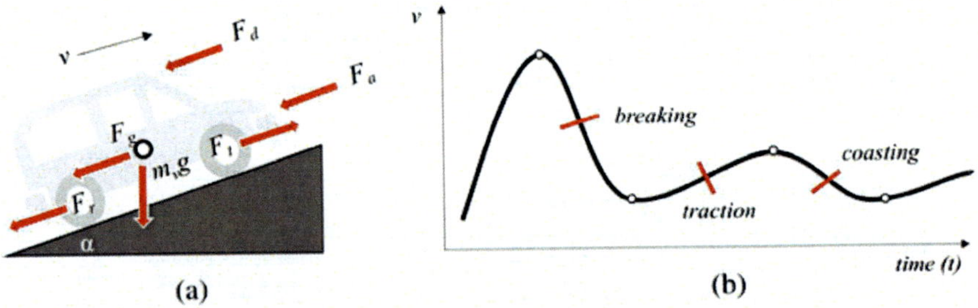

Fig. 2.10 **a** abstraction of the forces acting on a vehicle while moving, **b** modes of vehicle motion. Based on (Guzzella and Sciarretta 2013)

This dependency can be expressed with the following elementary equation (Eq. 2.3):

$$m_v \frac{d}{dt} v(t) = F_t(t) - F_a(t) + F_r(t) + F_g(t) + F_d(t) \tag{2.3}$$

where F_a is the aerodynamic force, F_r the rolling friction, F_g the gravity force, F_d is the disturbance force that entails all other forces not modelled in detail, and $F_t(t)$ is the total traction force generated by the vehicle to move minus the friction losses in the power train (Guzzella and Sciarretta 2013). Generally speaking, these forces are classified into conservative and dissipative forces. The most common dissipative forces are F_a and F_r. The aerodynamic force F_a is created by the friction of the air in the surroundings of the vehicle. It is normally estimated by simplifying the geometry of the vehicle as a prismatic body with a frontal area A_f. F_a is then expressed as:

$$F_a = \frac{1}{2} \cdot \rho_a \cdot A_f \cdot c_d \cdot v^2 \tag{2.4}$$

where v is the vehicle's speed, ρ_a is the density of the air at ambient temperature, and c_d is the aerodynamic drag coefficient which depends on the vehicle speed and wind flow direction. F_r is in turn expressed as:

$$F_r = c_r \cdot m \cdot g \cdot \cos(\alpha) \tag{2.5}$$

where m is the mass of the vehicle and c_r is the rolling resistance coefficient, which depends on the vehicle speed, road surface and tire pressure. The most relevant conservative forces are the gravitational force, F_g, which can be expressed as:

$$F_g = m \cdot g \cdot \cos(\alpha) \tag{2.6}$$

and the inertial (kinetic) force, F_k, caused by the acceleration and deceleration of the vehicle. F_k can be expressed as:

$$F_k = m_e \cdot a \tag{2.7}$$

where m_e is the equivalent mass of the vehicle and a its acceleration. The sum of these forces (Eqs. 2.4–2.7) is the traction force F_t (sometimes called force at the wheel). With F_t it is possible to express the power at the wheel P_w at a specific moment as:

$$P_w = F_t \cdot v \tag{2.8}$$

Depending on the value of P_w three operation modes can be identified as shown in Fig. 2.10b: (i) braking and/or regeneration ($P_w < 0$), meaning that the conservative forces have higher negative values than the dissipative forces have positive values. Under this operation mode the vehicles brakes and, if available, the recuperation device will absorb the negative torque, (ii) Traction ($P_w > 0$), the electric motor produces a positive torque to move the vehicle and (iii) idling or coasting ($P_w = 0$), conservative and dissipative forces cancelled each other out (e.g. idling or coasting) (Guzzella and Sciarretta 2013; Hofer 2014). Once P_w is known, the power that is taken from the battery at a given time point, $P_{bat,out}$, can be expressed as:

$$P_{bat,out} = \frac{P_w}{\gamma_{bat,disc} \cdot \gamma_{pe} \cdot \gamma_{em} \cdot \gamma_t} \tag{2.9}$$

where $\gamma_{bat,disc}$ is the discharge efficiency of the battery, γ_{pe} the efficiency of the power electronics, γ_{em} the efficiency of the electric motor and γ_t the efficiency of the transmission. Some reference values for these efficiencies are given in Table 2.1. An overview of these components is given in the following section of this chapter.

The mechanical energy demand for a given driving cycle can be expressed as the integral of P_w throughout the whole cycle and a distance x_{tot}:

$$E_m = \frac{1}{x_{tot}} \int_{t=0}^{T} P_{bat,out} dt \tag{2.10}$$

Table 2.1 Overview of efficiency ranges for selected current components in electric vehicles (Tschöke 2015)

Component	Efficiency	
	Min (%)	Max (%)
Battery	92	96
Power electronics	95	97
Transmission	80	95
Differential	92	98
Motor	87	95
Gearbox with fixed transmission	93	98

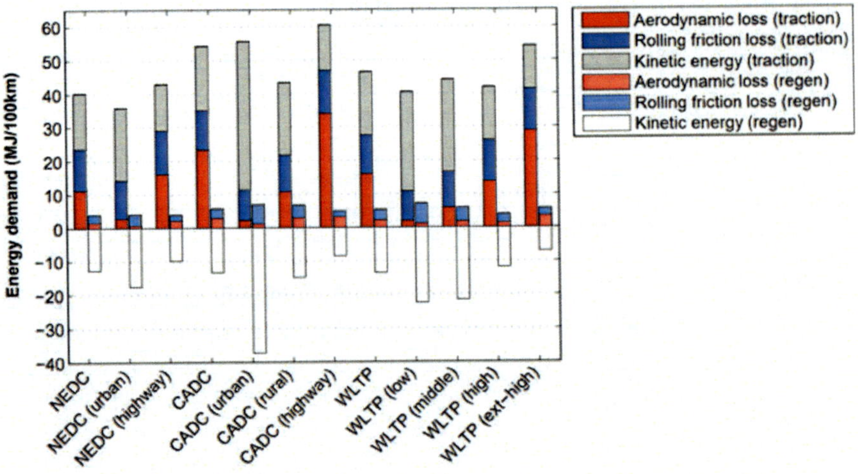

Fig. 2.11 Mechanical energy demand of an average European vehicle for different driving cycles (Hofer 2014)

A driving cycle is defined as a given schedule of the operation of vehicle as a function of speed and gear selection as a function of time (Barlow et al. 2009). The objective of a driving cycle is to provide a common base for the assessment of a vehicle's performance, for instance in terms of emissions, fuel consumed and drive train durability.

Some examples[20] of driving cycles are the New European Driving Cycle (NEDC), the Common Artemis Driving Cycle (CADC) and the Worldwide Harmonized Light Vehicles Test Procedure (WLTP). The WLTP is considered the most realistic of the driving cycles as it was design to represent real and modern driving conditions, and accordingly moves away from laboratory tests cycles. It is composed of four sub cycles: (i) low (simulates urban driving up to 56.5 km/h), (ii) medium (simulates suburban driving up to 76.6 km/h), (iii) high (simulates rural driving up to 97.4 km/h) and (iv) extra-high (simulates highway driving up to 131.3 km/h). Hofer (Hofer 2014) calculated the mechanical energy demand of an average sized European vehicle for several driving cycles, applying the equation given in 2.10. The results are presented in Fig. 2.11, which shows the contribution of the different forces to the energy consumption of the vehicle for the driving cycle.

This was accomplished by separating Eq. 2.10 into all the terms that compose P_w, which resulted in some important information. Considering only the results of the WLTP, it is noticeable that the urban cycles (low and middle) resulted in a higher share of recoverable energy than the rural and highway cycles. This is due to the fact that the urban cycles include a larger number of acceleration and de-acceleration events when compared to the cycles where higher speeds dominate. On the other side, the aerodynamic force is much higher in the rural and highway cycles, which

[20] A more detailed description of these and other driving cycles can be found in (Barlow et al. 2009).

Table 2.2 Power demand of selected auxiliary systems of vehicles (Duce et al. 2013)

Component	Mean electrical power [W]	Mean use ratio [%]
Lighting	140	75
Radio/navigation system	20	75
Seat heating (per seat)	30	5

means that more deceleration power is needed just to overcome the aerodynamic drag.

In addition to the mechanical energy required, energy for the heating, ventilation and air conditioning (HVAC) system, as well as for the auxiliaries needs to be accounted for. According to the eLCAr guidelines (Duce et al. 2013) the power demanded by the HVAC system ranges from 0.5 to 5 kW, depending on the outside temperatures, the efficiency of the devices and the users' profiles. There are different methodologies to evaluate or measure the consumption of these systems [see for example (Duce et al. 2013; Kambly and Bradley 2014; Egede 2016)]. Values for the most important devices are presented in Table 2.2.

2.2.3 Electric Vehicle Main Components

A higher electric-powered functionality increases an EV's size and weight requirements, as well as the complexity of its typical components (Guzzella and Sciarretta 2013). Regardless of the type of EV, a common electric drive train architecture (shown in Fig. 2.12) generally comprises an electric machine with a set of power electronic devices, including a charging system and a traction battery (Hofmann 2010; Herrmann and Rothfuss 2015; Wei Liu 2017).

An electric motor (EM) is an essential component of an EV. EMs can accomplish one or more of the following functions: (i) covert electric power into mechanical power, (ii) convert mechanical power from the internal combustion engine into electric power to charge the battery system and (iii) recover mechanical power from the drive train (Chan and Chau 1997; Guzzella and Sciarretta 2013; Un-noor et al. 2017). EMs are further classified into direct current (DC) and alternating current (AC). DC motors are considered to be simpler as they have required less of electronic devices to convert or invert current, and DC motors present disadvantages such as high maintenance and low efficiency when compared to AC motors. AC motors, in turn, are less expensive but their reliance on power electronics makes the whole electric drive system more expensive (Guzzella and Sciarretta 2013).

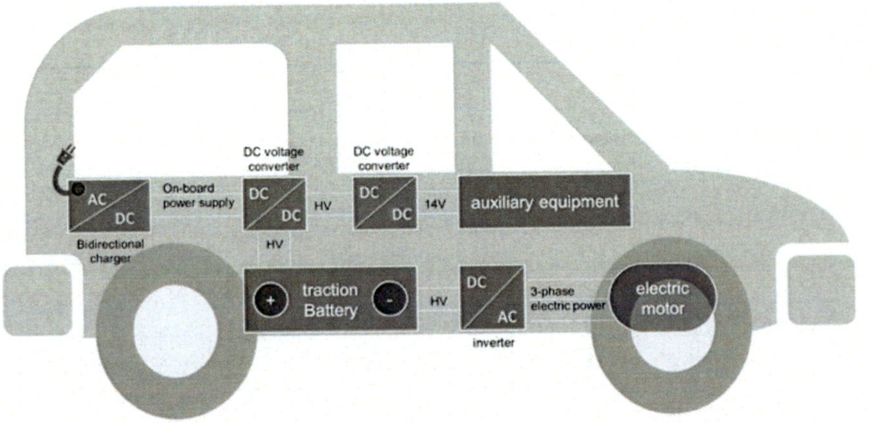

Fig. 2.12 Example of an electric drive train architecture. Based on (Hofmann 2010; Mi et al. 2011; Herrmann and Rothfuss 2015)[21]

[21] A thorough explanation and description of the most relevant components required by EVs is given in (Hofmann 2010; Guzzella and Sciarretta 2013; Tschöke 2015; Un-noor et al. 2017; Liu 2017).

Permanent magnet synchronous machines (PMSM) are one of the most common EMs within the commercially available EV models[22] (Grunditz and Thiringer 2016).This is due to the fact that PMSM present higher efficiencies compared to other EMs, have higher torque to current and torque to volume ratios, are smaller, more compact and respond faster (Islam et al. 2009). In a PMSM the stator has copper windings and the rotor has the permanent magnets (PMs) (Islam et al. 2009).

The typical material composition of PMSM has been modelled by Nordelöf and colleagues (Nordelöf et al. 2018). As reported, the most important materials in a PMSM are electrical steel (41–47%), aluminum (31–34%), copper (8–14%), carbon steel (4–8%) and neodymium magnets (2–3%) (Nordelöf et al. 2018).

Power electronics play a fundamental role in EVs as they invert and convert power depending on the requirements of the operating cycle. Typical devices used in EVs are: (i) DC-DC converters to convert the current voltage coming from the charger to the battery system and from the battery system to the different auxiliaries on board. (ii) DC-AC inverter to invert DC voltage from the battery system to be supplied to the EM. The process can be reversed in the case of regenerative braking, (iii) other basic elements such as diodes, thyristors and transistors (Mi et al. 2011; Liu 2017). To charge the battery system of an EV, a charger is required. Three levels of charging are defined by the electric power research institute (EPRI) as summarized in Table 2.3. Chargers present the same working principle and components configuration among

[22] Grunditz and colleagues (Grunditz and Thiringer 2016) present a detailed overview of current commercial EV regarding the type of EM and their influence on the vehicle's performance.

Table 2.3 Defined charging levels for EV (Liu 2017)

Level	Power capability	On board/off board
Level 1	Up to 1.5 kW	On board
Level 2	Up to 9.6 kW	On and off board
Level 3	50 to 150 kW	Off board

the three charging levels. They consist of an electromagnetic interface filter, a full wave rectifier, power factor correction (PFC) device and a DC-DC converter.

The traction battery of an EV is both the backbone and one of the biggest challenges of electromobility as these devices represent a significant part of the cost, vehicle mass and environmental impact of an electric vehicle. A traction battery is an energy storage device that provides the power and energy required for the propulsion of the vehicle. These batteries differ from starting, lighting and ignition (SLI) batteries in their operation requirements. Traction batteries are designed to be small and light, low-cost and with a reduced environmental footprint while offering high power-to-weight ratios, specific energy and energy density.

A battery is an ensemble of electrochemical cells (EC) connected in series or in parallel (Hambitzer et al. 2007). An EC is a device capable of generating electricity from a chemical reaction or of generating a chemical reaction using electricity. The ECs that generate electricity are usually called galvanic cell or voltaic cell. In an EC, the movement of electrons through an external circuit is produced by electrochemical processes taking place in within the electrodes of the cell. These electrochemical processes generate an internal movement of ions flowing through an electrolyte from one electrode to the other. Many types of EC exist, varying based on the nature of the chemical reaction and their design (Hambitzer et al. 2007). ECs are able to store and supply power over a given period of time (Goodenough 1998). In general, this means that a battery serves the basic functions of storing and converting energy (Goodenough 1998; Rahimzei et al. 2015). Regardless of the type of material contained in the battery cell, these are usually classified according to their principle of operation into fuel cells, primary cells and secondary cells (Hambitzer et al. 2007; Nishio and Furukawa 2007; Ketterer et al. 2009; Korthauer 2013; Kampker 2014; Rahimzei et al. 2015). Fuels cells (FC) convert the chemical energy from a reactant fuel continuously into electrical energy. The most common reactants for this application are hydrogen and oxygen which are required to be supplied from an external source. Contrary to FCs, the reactants in primary and secondary cells are contained within the cell. In primary batteries, the reacting compounds are completely consumed during the electrochemical reaction, thereby making it irreversible (Hambitzer et al. 2007). This means that this type of battery can only be charged and discharged once, and must be disposed of afterwards (Rahimzei et al. 2015). The secondary battery cells are rechargeable. In these cells, the application of external electrical energy reverses the electrochemical process, enabling the battery to store and supply energy over several cycles (Hambitzer et al. 2007; Ketterer et al. 2009). Lithium-ion batteries (LIB) are currently the most common type of batteries used in EV and are expected to continue dominating in relevance over the coming years. This thesis and the development of

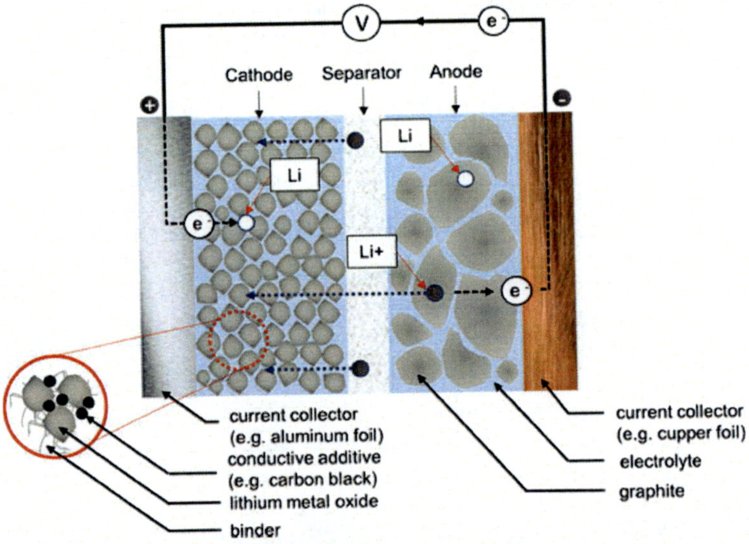

Fig. 2.13 Exemplary composition of a lithium-ion battery cell and abstraction of its operation principle during discharging. Based on (Korthauer 2013)

the concept presented in the following chapters is centered on this specific battery technology. Accordingly, the following section presents a more detailed overview of these battery technology.

2.2.4 *Lithium-Ion Traction Batteries*

Working principles and main cell components

A lithium-ion EC is composed of alternating cathode–anode-separator layers forming a stack. These layers are composed by positive electrodes (cathode) and negative electrodes (anode) (Korthauer 2013).[23] An abstraction of an exemplary stack of a lithium-ion cell is presented in Fig. 2.13. The electrodes consist of an active material coated over a metallic foil that serves as a current collector. These are normally copper (Cu) for the anode and Aluminum (Al) for the cathode, due to their corrosion resistance and high conductivity (Korthauer 2013). Between the electrodes, the cell contains an electrolyte, which enables the conduction of ions, and a porous membrane called separator, that isolates the electrodes from each other while at the same time permitting the movement of ions. In general, separator and electrolyte design deals with the trade-offs occurring between permeability properties,

[23] This classification of electrodes into cathode and anode is in reality only valid during the discharging phase of the battery. During charging, the positive electrode operates as the anode and the negative as the cathode.

ion conductivity, safety and mechanical strength (Kwade et al. 2018). The electrolyte in a lithium-ion battery consist of a lithium salt such as lithium hexafluorophosphate $LiPF_6$ or lithium tetrafluoroborate ($LiBF_4$) disassociated in an organic carbonate solvent (Besenhard 2007; Korthauer 2013). The mixture of carbonate solvents such as ethylene carbonate (EC), ethyl methyl carbonate (EMC and vinylene carbonate (VC), enables the formation of a protective film at the interphase between the electrolyte and graphite particles called the solid electrolyte interphase (SEI). The lithium salt enhances ion conductivity (Kwade et al. 2018).

The separator is an essential part of the cell, as it can strongly influence the functioning of a cell. Among the most important requirements are permeability, porosity, thickness and thermal shutdown. In lithium-ion batteries, the most common materials used as separators are microporous membranes made from PE, polypropylene (PP) or polytetrafluoroethylene (PTFE). In larger cell formats, this polymer layer is coated with layers of ceramic materials such as Al_2CO_3 and SiO_2 for safety reasons (Kwade et al. 2018). Additional ceramic coating serves as an electrical isolation barrier and decreases the risk of heating up during cycling.

In a lithium-ion battery cell, both a discharging and a charging process take place. During the discharging process, the active material at the anode is oxidized due to the electric potential difference between the two electrodes. This causes lithium-ions to dissociate and to move from the negative electrode through the electrolyte and the separator to the positive electrode. Electrons are conducted simultaneously through the external circuit, providing usable electrical energy.

This movement of electrons through the external circuit and ions through the electrolyte is inverted during the charging process. Here, the application of an external electricity supply forces the electrons to leave the cathode and move to the anode while internally the lithium-ions move in the same direction (Besenhard 2007; Korthauer 2013; Deng 2015). In this way, the electrical energy is stored as chemical energy. The potential difference between the two electrodes is estimated as the difference between two half cells. This implies that the potential of a galvanic cell is given by a pair of electrodes (Besenhard 2007), and so, in order to design a cell that can be used as a source of power, the pair of electrodes are chosen in such a way that their potentials are as far from one another as possible in the electrochemical series[24] (Besenhard 2007). In the exemplary cell configuration from Fig. 2.13, graphite as the anode is combined with a lithium metal oxide cathode. In general, carbonaceous materials such as natural and synthetic graphite, as well as amorphous carbons, are currently the most commonly used anode materials, accounting for 96% of the whole market in 2016 (Schmuch et al. 2018). Additionally, graphite and amorphous carbons are frequently combined with the objective of optimizing the power to energy ratio, as reviewed in (Schmuch et al. 2018). Other technically relevant anode materials are Silicon/Graphite composite (Si–C), lithium titanate (LTO) and lithium metal. LTO is known to offer an improved cycle stability and a high safety performance. Nevertheless, its specific energy and its potential (vs. Li/Li+) is very high, resulting in a low cell voltage (Korthauer 2013; Schmuch et al. 2018). Although

[24] Arrangement of elements based on the increasing reduction potential values.

Si–C composites are still on a laboratory level, some important challenges have been acknowledged in regard to its change of volume and the related mechanical degradation and loss of lithium (Nitta and Yushin 2014; Nitta et al. 2014; Schmuch et al. 2018). Lithium metal is currently seen as one of the most promising anode materials due to its high specific energy and energy density, particularly in combination with solid electrolytes. Nevertheless, this technology still faces important challenges regarding its safety during operation and the manufacturability of the lithium metal foil, which in turn influences its production costs as well as safety issues during production (Korthauer 2013; Schmuch et al. 2018).

The most important requirements of the positive electrode materials include high specific energy and energy density, high discharge potentials (vs. Li/Li+), safety, low costs and good capacity retention (Besenhard 2007; Korthauer 2013; Schmuch et al. 2018).

There are many possible metals that can be used as active materials for the positive electrode (Placke et al. 2017). These can be classified according their crystal structure into: layered oxides, spinel and phosphate (Korthauer 2013).

$LiMO_2$-type (M=Co, and/or Ni and/or Mn) layered oxide cathodes are currently the most researched materials. This type of materials include transition metals (M) such as cobalt (LCO); nickel (LNO); nickel, cobalt and aluminum (NCA); manganese (LMO); nickel and manganese (NMO); or nickel, manganese and cobalt (NMC). LCO cells is known as the first layered oxide cathode to have been commercialized (Nitta et al. 2014). It offers a relatively high theoretical specific energy and energy density, good cycling performance and high discharge voltage (Besenhard 2007; Nitta et al. 2014). However, it faces some significant limitations, such as low thermal stability, which can lead to a thermal runaway (it has the lowest among any current commercial cathode material), accelerated capacity fade at high current rates, and high costs due to the its high contents of cobalt (Nitta et al. 2014). LNO has lower costs compared to LCO and presents a similar specific energy, however these cathodes are seen as not favorable since Ni^{2+} have been found to substitute lithium-ions during de-lithiation and therefore block the Li diffusion process (Nitta et al. 2014). A further strategy identified to improve thermal stability and electrochemical performance is the addition of aluminum (Al). In this regard, the active material $LiNi_{0,8}Co_{0,15}Al_{0,05}$ (NCA) has been found to reach around 200 $mAhg^{-1}$ and exhibit a longer calendar life compared to LCO, which enabled its widespread commercial application. Since Mn is cheaper and less toxic than Co or Ni, $LiMnO_2$, has been seen as a promising alternative. However, some challenges regarding its cycle life have been identified, mainly that Mn tends to leach out of LMO during cycling (Nitta et al. 2014). $Li(Ni_{0,5}Mn_{0,5})O_2$ (NMO) has been seen as an attractive material, as it could achieve a comparable specific energy as the LCO, but with a reduced material cost due to the content reduction of cobalt. Its rate capability has been found to be very low (Nitta et al. 2014), and so, a further strategy to improve the structure stability was found to be the addition of Co into the $Li(Ni_{0,5}Mn_{0,5})O$. The resulting $LiNi_xCo_yMn_zO_2$ (NMC) enabled the achievement of higher specific energy values than LCO but with much lower costs, due to the reduced content of cobalt. Since a higher share of Ni results in higher capacity. This strategy has been adopted in the development of improved cathode materials. As reported by Schmuch and colleagues (Schmuch et al. 2018), an increase in capacity from 160 $mAhg^{-1}$ for $LiNi_{0,33}Co_{0,33}Mn_{0,33}O_2$ (NMC-111) to up to 200 $mAhg^{-1}$ $LiNi_{0,8}Co_{0,1}Mn_{0,1}O_2$ (NMC-811) can be achieved. However, this increase in nickel comes along with a reduction of lifetime, safety and thermal stability.

The most important spinel type cathode material is $Li_{1-x}Mn_2O_4$ (also called LMO or LMO-spinel). As is implied by its name, LMO-spinel has a 3-dimensional spinel structure that enables a higher rate capability because it provides a diffusion path for the lithium ions easing their intercalation and de-intercalation processes (Besenhard 2007). Moreover, as this material does not contain cobalt or nickel, it presents significant advantages regarding costs and environmental issues (Nitta et al. 2014; Ellingsen et al. 2016). Nevertheless, LMO-spinel has been found to face some challenges regarding its cyclability (Nitta et al. 2014).

Finally, the phosphate group $LiMPO_4$ (M=Fe, Mn, Co, Ni) consists of cathode materials with olivine structures. In this group, $LiFePO_4$ (LFP) is the most widespread material in commercial application due to its very good thermal stability and high power capabilities. Major weaknesses of LFP are its low potential as well as low electrical and ionic conductivity (Nitta et al. 2014).

A comparison of the different cathode materials previously reviewed is shown in Fig. 2.14. As seen, layered oxide materials present a clear advantage when it comes to specific energy. Notice that although LFP does not contain expensive materials such as nickel, cobalt or manganese, the cost on a kWh basis is comparable to the case of the layered oxide metals due to its lower specific energy. Nevertheless, in

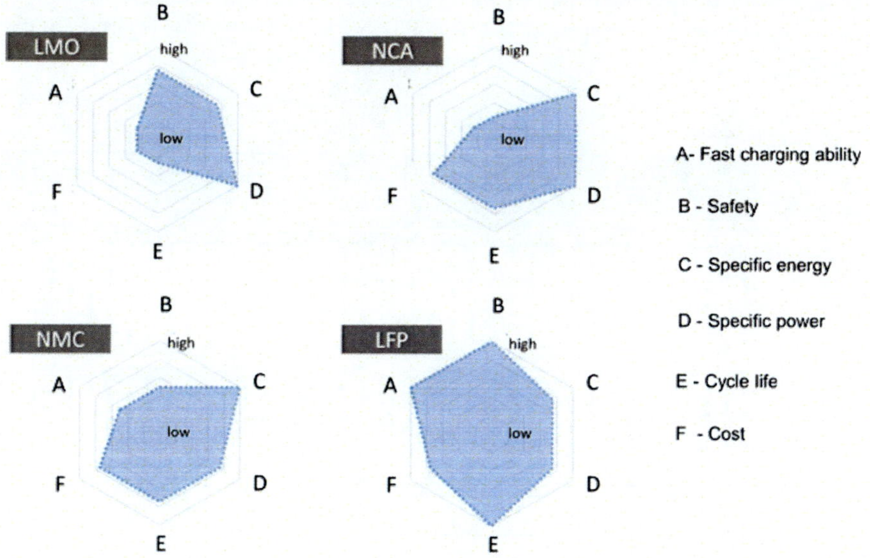

Fig. 2.14 Comparison of different cathode materials regarding different considerations. Adapted from (Korthauer 2013)

terms of safety issues LFP presents a significant advantage compared to the rest of the materials as a thermal runaway in a LFP cell is very unlikely to happen when compared to the layered oxide materials (Korthauer 2013).

In conclusion, there is a variety of cathode materials, each with different properties and requirements that influence the type of application. While layered oxide batteries have currently a dominant position in the BEV market due to its specific energy and power abilities, LFP batteries have been found to be reliable in applications where safe and long-lasting cathode materials are desirable such as some HEV, trucks and buses as well as stationary applications.

Battery design and manufacturing

An abstraction of the configuration of a battery pack is shown in Fig. 2.15. A battery system consists of modules of cells connected in parallel or in serial which leads to higher voltage or capacity respectively.

The cells in the modules are also connected in serial or parallel depending on the system requirements. Additionally, the battery pack contains a Battery Management Unit (BMU, also called battery management system BMS), a thermal management system (TMS) and a set of power electronics (PE) (Kampker et al. 2013b). A BMU is made up of electronic components and software which primarily functions to protect the cells in order to extend their life cycle and number of cycles (Dorn et al. 2013). The most important control parameters of the BMU are the cell voltage, cell temperature and current. Additionally, through the integration of electrochemistry and electrical models, along with statistical methodologies, the BMU estimates the State of Charge (SOC) and the State of Health (SOH) of the battery system. SOC is a metric of the actual charge level of the battery pack and SOH describes the condition of the battery compared to a new one (Dorn et al. 2013). The function of the TMS is

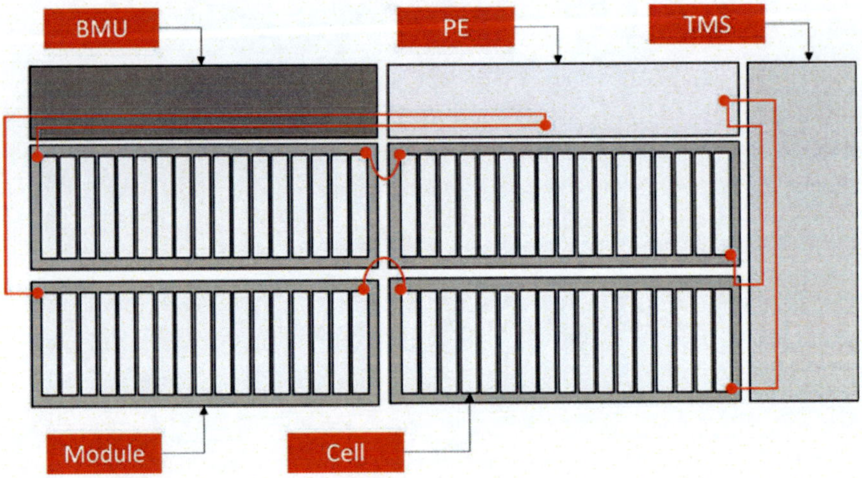

Fig. 2.15 Representation of a configuration of a battery pack

to keep the operating temperature of the battery system within a safe range, which prolongs its life time (Zeyen and Wiebelt 2013).

The optimal operation temperature of the battery pack of a lithium ion battery system ranges from 20 to 40 °C, due to both the good conductivity properties and aging mechanism shown by li-ion ECs. Thus, both cooling and heating systems are necessary to assure the proper functioning of the traction battery. Cooling concepts currently applied to lithium ion battery packs include the: (i) air cooling system, (ii) direct expansion fluid cooling system and (iii) indirect (secondary circuit) cooling system. Some heating concepts applied to traction batteries are the: (i) application of inert liquids, (ii) direct heating through an electrical heating element and (iii) indirect electric heating element.

Disregarding the cell chemistry used for the electrodes, battery cells can be classified both according to their geometrical form and by characteristics of their housing. In this regard, the most common formats of traction battery cells commercially available are cylindrical, pouch and prismatic (Michaelis et al. 2018).

A comparison of the relevant properties of the different cell geometries are summarized in Table 2.4. Pouch and prismatic cells are geometrically similar as both formats present a prismatic volume. This type of geometry offers a more efficient use of the volume available for electrodes, which leads to a higher packaging density inside the cell. The casing material of a pouch cell is typically a multi-layer aluminum foil (also called soft case), while the prismatic cells are predominantly contained in hard aluminum cases. The hard case prismatic cell typically presents a lower specific and

Table 2.4 Comparison of characteristics of common cell formats (Kampker et al. 2013a, b; Michaelis et al. 2018)

Category	Pouch cell	Prismatic cell	Cylindrical cell
Energy density at cell level	+++	++	+++
Energy density at module level	+++	+	+++
Influence of geometry	– High cell design flexibility – Efficient use of space – High packaging density	– Low cell design flexibility – Efficient use of space – High packaging density	– Low cell design flexibility – Inefficient use of space – High packaging density
Mechanical characteristics	– Instable case – Swells under increased pressure	– High stiffness – Robust package – Keep its form under increased pressure	– High stiffness – Robust package – Keep its form under increased pressure
Thermal characteristics	– Good surface-volume relation which enables a better heat dissipation	– Less optimal surface volume relation which leads to a less efficient heat dissipation	– Inefficient heat dissipation

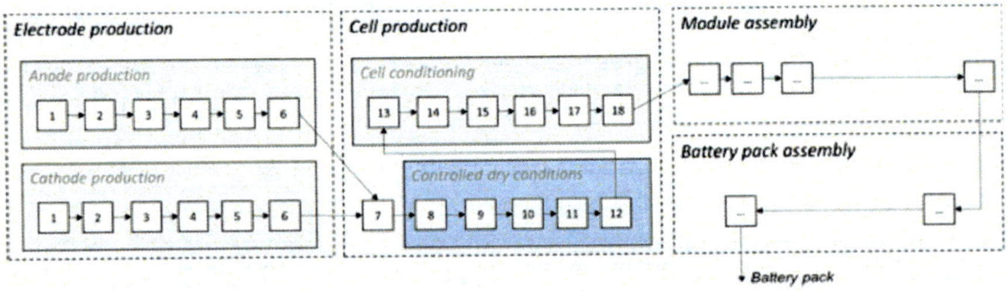

Fig. 2.16 Process chain in the production of lithium-ion traction batteries (Gallagher and Nelson 2014; Kampker 2014; Kwade et al. 2018)

volumetric energy compared to the pouch cell, due to the difference of weight in both cells. Cylindrical cells have been estimated to be more inefficient in terms of volume usability compared to prismatic and pouch cells. However, their casing is composed of nickel-plated-steel which is light enough to enable the cell to achieve a similar or even higher specific energy compared to the pouch cell (Michaelis et al. 2018).

The manufacturing process of a traction battery can be classified into three main phases as represented in Fig. 2.16: (i) electrode production, (ii) cell production, (iii) module and pack assembly. In the electrode production process, with the exception of the material preparation step, the process chain for the production of anodes and cathodes consist of the same processes. The first step (1) is called *dry and wet mixing* (Kwade et al. 2018). The dry mixing process can be executed through a dry homogenization or an intensive dry mixing. This step has the objective to mix the powdery components such as the active cathode or anode materials with the conductive additive (e.g. carbon black). The dry mixing process has been proven to improve the homogeneity of the resulting slurry which subsequently leads to an enhanced cycling stability (Bockholt et al. 2016). The following wet mixing process uses a high intensity mixer, such as planetary mixer, to disperse the dry mixture with further additives such as solvents (e.g. N-methylpyrrolidone NMP) as a viscosity regulator and the binder (commonly polyvinylidene fluoride PVDF). The result of this mixing process is a slurry mixture of active materials with a particular viscosity.

The next step (2) is called *coating*. Here the mixture is coated on the current collector, typically through the application of a slot die coating machine. The slurry might be coated continuously or intermittently on one or both sides of the current collector, depending on the machinery used. A simultaneous double coating machine is nevertheless not very common in industrial processes due to concerns regarding potential structural variations among the two sides (Kwade et al. 2018).

Immediately after the coating process, a *drying* process (3) is applied, with the objective of evaporating the solvent from the slurry, leading to a solidification of the coated layer (Kaiser et al. 2014). The drying process is very critical, as it represents a significant share of the total cell manufacturing costs due to the energy required. Furthermore, parameters such as temperature or drying-speed have been demonstrated to have a important influence on the final cell performance, as they affect the adhesion of the coating layer to the current collector (Kaiser et al. 2014; Schönemann 2017a).

The resulting intermediate product presents a high porosity meaning a low density, low adhesion and low conductivity. The subsequently *calendering* process (4) has the objective of compressing the coated material to gradually reduce the thickness of the coated layer. This is done by passing the electrode through a two-roller calender with typical roll diameters of 600–1000 mm at speeds between 30 and 100 m min^{-1} (Kaiser et al. 2014; Meyer et al. 2017; Kwade et al. 2018). The process not only makes the coated layer more compact, improving its density and conductivity, but also enhances the contact between the current collector and the coated layer.

After the calendering processes, the manufactured electrodes are *slitted to width* (5) by applying a continuous process using laser or knifes (Kampker 2014) and finally dried (6) to eliminate any water content that still remains in the coated layer.

The cell production process starts with a *cutting process* (7) in which, depending on the type and geometry of cell being manufactured, the electrodes are precisely cut to meet tolerances. Then, the cut electrodes and separator are transferred to a dry room with controlled environmental conditions (temperature 20 °C and dew point between −40 and −60 °C) (Kwade et al. 2018). The next step is called cell *packaging* (8). The objective here is to place the so-called electrode-separator assembly (ESA) into a cell housing (Kwade et al. 2018). This process depends on the type and geometry of cell being manufactured (see Table 2.4). Three type of processes are identified as the most common to fulfill this task: The first is a winding process which is normally applied for the assembly of small cells in cylindrical or prismatic format (Kampker 2014; Schmitt et al. 2014). The second type is a stacking process in which the individual layers of separator, cathode and anode are stacked over each other following the order cathode-separator-anode-separator (Schmitt et al. 2014). The third type is the so-called Z-folding process in which a separator is continuously fed and folded into a z-shape to which the individual electrode films are discretely inserted, forming an ESA (Kampker 2014; Schmitt et al. 2014; Schilling et al. 2016; Schönemann 2017b; Kwade et al. 2018). In the next two steps, *contacting terminals* (9) and housing *and sealing* (10) the ESA is connected internally through welding the tabs, and then placed into a cell housing that can be a hard case or an aluminum pouch-bag as summarized in Table 2.4. The last two steps of the cell assembly process that take place inside the dry room are the *electrolyte filling* process (11) and the *final sealing* of the cell (12). The electrolyte filling process takes place under rigorous ambient conditions, including a dew point lower than −60 °C at a temperature of 20 °C and a weak vacuum. The process is very time consuming as the electrolyte must filled all porous cavities within the ESAs by displacing the air trapped in its structure. Once the cell is filled, the housing is sealed and stored under controlled ambient temperature conditions (Schmitt et al. 2014). The cell conditioning step is composed of a formation process (14), an aging process (16) and three *quality control* steps [*initial* (13), *intermittent* (15) and *final* (17)]. The formation process is basically the first charging of the manufactured cell, (Schönemann 2017b) and has been identified as one of the largest contributors to the manufacturing costs of lithium-ion battery cells (An et al. 2017). The purpose of this step is to activate the active materials and start forming the SEI at the anode. The SEI is formed mostly during the first charging-discharging cycle, when the electrons at the electrodes access the electrolyte. The electrolyte precipitates at the anode via reduction reactions, forming a passivation layer that isolates the electrode from the electrolyte, preventing an irreversible consumption of lithium-ions (An et al. 2017). The formation process is built within one day, and is followed by an aging process in which further charging and discharging cycles are executed to the cell at particular amperage, pause lengths and temperatures (Schönemann 2017b). This process normally takes between two and three weeks and represent up to 30% of the investment costs (Kwade et al. 2018).

Finished battery cells are subsequently arranged into modules as represented in Fig. 2.16. Finally, the modules are assembled into a battery pack. The assembly process chain of a battery pack is constrained by aspects of product design, but a standard chain of processes is nevertheless described by (VDMA 2012).

Battery degradation

A representation of the most common degradation mechanisms, their cause effect links and their modes is shown in Fig. 2.17, provided by Birkl and colleagues (Birkl et al. 2017).

A lithium-ion cell degrades due to its usage and its exposure to ambient conditions. As a consequence, the ability of the cell to store energy and supply a particular power demand is affected, marking the end of the cell's useful time (Birkl et al. 2017). In the figure, the authors summarized the most commonly reported modes of degradation suffered by a cell. As shown, the three most relevant modes are: (i) the loss of lithium inventory due to mechanisms that lead to an irreversible consumption of lithium such as SEI growth, SEI decomposition, electrolyte decomposition or lithium plating and

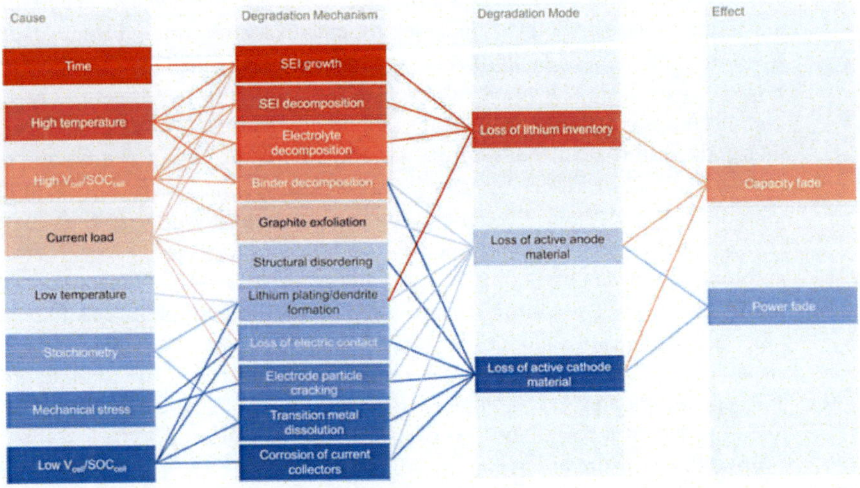

Fig. 2.17 Cause-effect pathway of the degradation mechanisms acting on lithium-ion cells. Taken from (Birkl et al. 2017)

dendrite formation, (ii) the loss of active anode materials and (iii) the loss of active cathode material.

Battery Recycling

Battery recycling plays an essential role in supporting the transition towards an electrified transportation sector. Recycling traction batteries is necessary, as it can reduce their life cycle environmental impact by avoiding the production of virgin materials, limiting the undesirable environmental impacts that rise from alternative E-o-L treatments (i.e. landfilling or incineration), and enabling the local production of materials that are otherwise geopolitically limited (Cerdas et al. 2018a).

The recycling processes for traction batteries face several technical, economic, environmental and safety challenges (like electrical and chemical hazards).

For instance, the high variability of chemistries, cell geometric formats and pack configurations (e.g. regarding additional components such as electronics and cooling systems), affects the efficiency of the designed process. There are currently a number of different possible recycling routes available for lithium-ion batteries composed of different combinations of unit process such as deactivation, pyrometallurgical treatment, mechanical separation steps and hydrometallurgical steps (Hanisch et al. 2015; Diekmann et al. 2017; Velázquez-Martínez et al. 2019). The deactivation is generally carried out by discharging the battery completely or through a thermal pre-treatment process, in which the battery is heated up in order for the electrolyte to evaporate (Diekmann et al. 2017). Another possible process to deactivate the battery system prior to its recycling processing is freezing (Hanisch et al. 2015). Here, the battery pack is frozen under −65 °C, preventing chemical reactions the transition of hazardous materials (Hanisch et al. 2015). Depending on the recycling path followed, a partial or full disassembly of the battery system into modules or even cells is usually manually done (Wegener et al. 2014; Cerdas et al. 2018d). Mechanical treatment consists of a sequence of processes that convert the battery pack into a flow of material mixture that is capable of being bulk stored and transported (Hanisch et al. 2015). Here the most common processes are crushing (wet and dry) and other classification and sorting processes (Diekmann et al. 2017). Pyrometallurgical processes recover metals from a batch of batteries with different cell chemistries, and can even be combined with production routes for virgin materials (Hanisch et al. 2015). Unfortunately, pyrometallurgical processes are normally linked to high energy consumption and present some limitations in that it is not possible to regain lithium or aluminum through these processes (Hanisch et al. 2015). Hydrometallurgical processes consist of a combination of leaching, extraction, crystallization and precipitation processes (Hanisch et al. 2015). These are applied to recover metals such as Co, Ni, Mn and Li from the mechanical treatment steps (Wang and Friedrich 2015; Diekmann et al. 2017). Additionally, Al and Li can be extracted from pyrometallurgical slags using hydrometallurgical processes (Diekmann et al. 2017).

2.3 Life Cycle Engineering of Battery Electric Vehicles

2.3.1 LCA of Electromobility

The environmental impact of electric vehicles is a topic that has been intensively discussed within the fields of science and politics. Important methodological progress has been made vis a vis this topic, as it has been recognized that considering the supply chain of components such as the battery is essential to execute a consistent comparison and to support policy making and strategic planning. The manufacturing and raw material stages in particular have increasingly gained attention. Efforts have been made to standardize the methodology in the field of electromobility. Examples of these efforts are the development of guidelines (Duce et al. 2013; Cerdas et al. 2018c; EEA 2018; Van Loon et al. 2018; Emilsson and Dahllöf 2019) and some more integrated methodologies (Bauer et al. 2015; Egede 2017; Cox et al. 2018), which have been advanced in the last decade, with the purpose of setting the ground work for more transparent modelling approaches. These methodological guidelines assist the LCA practitioner in consistently defining the goal and scope of an LCA study and its system boundaries in order to plan data collection processes. This has led to the emergence of several frameworks for the definition of system boundaries, such as the one shown in Fig. 2.18. While the system boundary for the evaluation of a conventional vehicle has historically been focused on well-to-wheel (WTW) and tank-to-wheel (TTW) approaches, with the diversification of power train technologies, the application of LCA for the assessment of EVs has become increasingly relevant. As opposed to ICEVs, whose environmental hotspots traditionally lie in the usage stage, the environmental hotspots of a EV are expected to shift to other stages

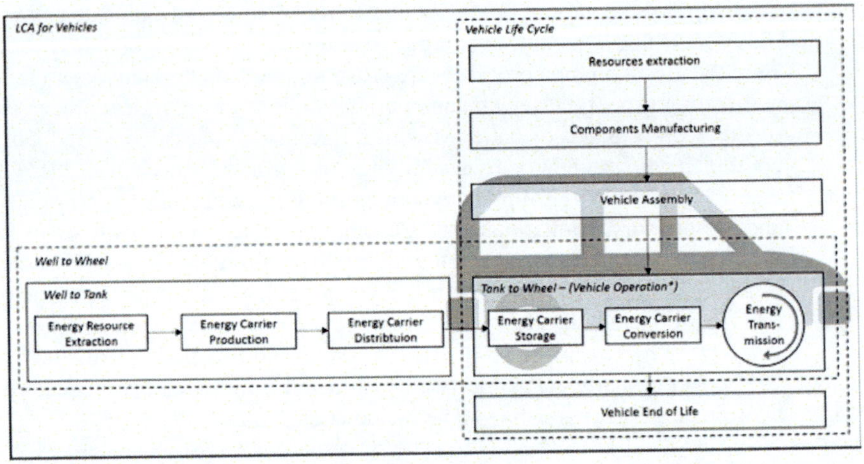

Fig. 2.18 System boundaries for the application of LCA for EVs (Nordelöf et al. 2014; Cerdas et al. 2018c)

such as manufacturing, raw material production and recycling. Thus, as the transition towards electromobility becomes more compelling, the EV's entire life cycle, including the production of materials and components, as well as the recycling phase, gains in relevance.

The aforementioned guidelines additionally provide structure to better plan data collection processes for the development of a life cycle inventory, and help to better identify the significant issues that must be considered throughout the life cycle stages.

Raw materials extraction, production and assembly

The significant issues identified in these life cycle stages by the guidelines are structured into the different components required by an EV, as shown in Fig. 2.20. Some aspects of the power train, such as the electric motor, the on-board power electronics (charging electronics and other devices such as the inverter and controller and capacitors) and the battery system, are classified as significant, due to their reliance on materials associated with environmentally intensive production chains (e.g. copper, nickel, cobalt, aluminum, lithium and graphite, as well as the use of rare metals for some magnets such as neodymium, and other critical materials like copper, gold, silver, tin and platinum) (Duce et al. 2013; Van Loon et al. 2018).

Additionally, the use of critical raw materials is seen as a significant issue when analyzing the environmental impact of an electric vehicle, as this use implies a potential supply constrain in the future due limited geopolitical availability.

An additional issue identify as significant within these life cycle stages is the potentially increased environmental impact linked to the energy consumption of battery cells during the manufacturing stage.

Electric vehicle usage stage

There are three significant issues that the guidelines highlight during the usage stage. First, driving and charging patterns are categorized as significant, as the energy used to power cabin services such as heat, ventilation and air conditioning potential increases. Other sources of energy consumption, such as efficiency losses in the transmission of energy throughout the charging process or due to standstill discharging, are also considered significant.

The way that the interaction of different elements of the EV impacts energy consumption is also considered a significant issue. Because of this, the eLCAr guidelines (Duce et al. 2013) introduced a methodology that aims to identify the interactions between the components (Fig. 2.19) and considers potential high order degree interactions between the components.

A second significant issue identified by the guidelines that occurs during the usage stage of an EVs life cycle is the maintenance processes, particularly with regard to the possibility of a battery replacement. And finally, the third issue is that of the non-tailpipe particulate emissions which are produced in the braking system.

Vehicle End-of-life

While this life cycle stage is usually neglected, the guidelines (Duce et al. 2013; EEA 2018; Van Loon et al. 2018) consider the EV's end-of-life stage to be relevant due to

	is/are influenced by … / influence(s)…	COMPONENTS: Charger	Traction Battery	Power Electronics	E-motor	Transmission	SBSS	Body (Frame)	Aerodynamics	Insulation	Non-p. El. System	VEHICLE: Vehicle Weight	Overall Energy Demand (in use phase)
VEHICLE	Vehicle Weight	1	1	1	1	1	1	1	0	1	1		1
	Energy Demand (in use phase, NOT via weight)	1	1	1	1	1	1	0	1	1	1	1	
COMPONENTS	Charger		1	0	0	0	0	0	0	0	0	0	0
	Traction Battery	0		1	1	0	0	0	0	0	0	0	1
	Power Electronics	0	1		1	0	0	0	0	0	0	0	1
	E-motor	0	0	0		0	0	0	0	0	0	0	1
	Transmission	0	0	0	1		0	0	0	0	0	0	1
	SBSS	0	0	0	0	0		0	0	0	0	1	1
	Body (Frame)	0	1	0	0	0	0		0	0	0	1	0
	Aerodynamics	0	0	0	0	0	0	0		0	0	0	0
	Insulation	0	0	0	0	0	0	0	0		0	0	0
	Non-p. El. System	0	0	0	0	0	0	0	0	1		0	0

Fig. 2.19 Interdependency matrix of EV components showing their effect on vehicle weight and overall energy demand (Duce et al. 2013)

the potential in this stage to recover the essential raw materials that can significantly reduce the environmental impact of EV in future products. What's more, recycling has the potential to minimize the burdens linked to other disposal treatments and to limit the interaction between hazardous materials and the biosphere.

2.3.2 Modelling Complexity

Understanding how electrical vehicles interact with the environment, in the context of discussions of sustainable development or climate change mitigation, involves utilizing a complex system. A complex system is by definition composed of a large number of entities that are highly interconnected, interrelated and integrated (Crawley et al. 2015). Ashby and colleagues further argue that the complexity of sustainability issues is especially challenging, because the problem is often characterized by being multi-dimensional, interactive and poorly defined (Ashby 2016).

Electric vehicles are highly interconnected and integrated technology, and their environmental footprint is greatly influenced by a variability of parameters regarding geographical and temporal aspects. Thus, modelling the life cycle of an EV is a complex task, as represented in Fig. 2.20. Each of the life cycle stages that are required to be considered presents a high parameter variability, defined by specific

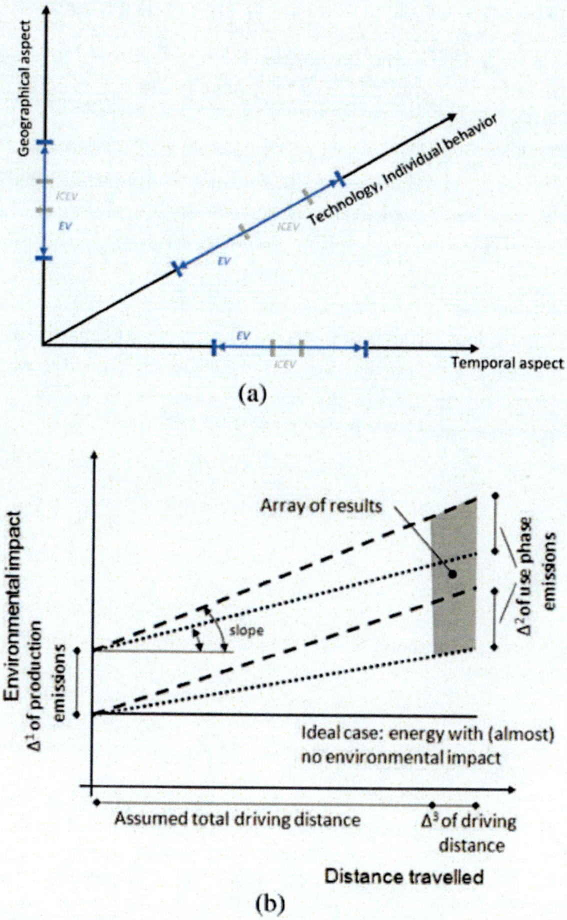

Fig. 2.20 Variability in the life cycle modelling of electric vehicles. **a** Variability in the modelling of life cycle inventories (Own figure.). **b** range of LCA results (Egede et al. 2015)

technological, geographical or temporal contexts. What's more, each implies particular interactions between foreground and background systems and between other life cycle stages (see Fig. 2.20a). Complicating the matter further, research in this field is extremely multidisciplinary, spanning a multitude of topics including the development of new active cathode materials for traction batteries; new coating processes; better recycling process; fleet composition; supply chain optimization; and business models, to name a few. This complexity has the natural repercussion of causing variability in the results, which can take values in large ranges, therefore challenging decision-making processes (see Fig. 2.20b).

While not erroneous, the results of currently available LCA research studies have raised important questions that haven't been clearly answered: (i) first, is it possible to argue that there is a clear environmental advantage between both technologies? (ii) what is the combined effect of different influential factors on the estimated GWP of electric vehicles? and (iii) how can these results be used to make consistent design and engineering decisions?

With this motivation, a significant number of LCAs for electric vehicles diverging in goal and scope have emerged in the last few years. Plenty of important research questions have been explored in order to better understand the questions mentioned above. But while these questions might be partially answered one LCA study at a time, the greater problem is fundamentally of methodological origin.

This is twofold problem. On the one hand, although a standard methodology (LCA) exist, it has important inherent challenges, as discussed in Sect. 2.1.3, making it less than ideal in engineering and decision making. On the other hand, while the methodology offers a scientific basis to support the engineering of EVs, the newer technology present much more complex system boundaries and interactions with other technological and societal systems than conventional ICE.

References

Abele E, Anderl R, Birkhofer H (2005) Environmentally-friendly product development. Springer, London, London

Alting PL, Jøgensen PJ (1993) The life cycle concept as a basis for sustainable industrial production. CIRP Ann Manuf Technol 42:163–167. https://doi.org/10.1016/S0007-8506(07)62417-2

Alting L, Legarth JB (1995) Life cycle engineering and design. Ann CIRP 44:569–580. https://doi.org/10.1016/S0007-8506(07)60504-6

An SJ, Li J, Du Z et al (2017) Fast formation cycling for lithium ion batteries. J Power Sources 342:846–852. https://doi.org/10.1016/j.jpowsour.2017.01.011

Ashby MF (2016) Materials and sustainable development. Elsevier

Ashby MF (2013) Materials and the environment eco-informed material choice

Ayres RU, Van Den BJCJM, Gowdy JM (2001) Strong versus weak sustainability: economics, natural sciences, and "consilience." Environ Ethics 23:155–168. https://doi.org/10.5840/enviroethics200123225

Baitz M, Albrecht S, Brauner E et al (2013) LCA's theory and practice: like ebony and ivory living in perfect harmony? Int J Life Cycle Assess 18:5–13. https://doi.org/10.1007/s11367-012-0476-x

Barlow T, Latham S, Mccrae I, Boulter P (2009) A reference book of driving cycles for use in the measurement of road vehicle emissions. 280

Bauer C, Hofer J, Althaus HJ et al (2015) The environmental performance of current and future passenger vehicles: life cycle assessment based on a novel scenario analysis framework. Appl Energy 157:871–883. https://doi.org/10.1016/j.apenergy.2015.01.019

Beltran AM, Prado V et al (2018) Quantified uncertainties in comparative life cycle assessment: what can be concluded? Environ Sci Technol

Besenhard JO (2007) Handbook of battery materials

Bey N, Hauschild MZ, McAloone TC (2013) Drivers and barriers for implementation of environmental strategies in manufacturing companies. CIRP Ann Manuf Technol 62:43–46. https://doi.org/10.1016/j.cirp.2013.03.001

Bhander GS, Hauschild M, McAloone T (2003) Implementing life cycle assessment in product development. Environ Prog 22:255–267. https://doi.org/10.1002/ep.670220414

Birkl CR, Roberts MR, Mcturk E et al (2017) Degradation diagnostics for lithium ion cells. J Power Sources 341:373–386. https://doi.org/10.1016/j.jpowsour.2016.12.011

Bjørn A, Owsianiak M, Laurent A et al (2018) Scope Definition. Life cycle assessment. Springer International Publishing, Cham, pp 75–116

Bjørn A, Owsianiak M, Molin C, Laurent A (2018) Main characteristics of LCA. Life cycle assessment. Springer International Publishing, Cham, pp 9–16

Bjørn A, Bey N, Georg S et al (2015) Is Earth recognized as a finite system in corporate responsibility reporting? J Clean Prod 1–12. https://doi.org/10.1016/j.jclepro.2015.12.095

Bockholt H, Haselrieder W, Kwade A (2016) Intensive powder mixing for dry dispersing of carbon black and its relevance for lithium-ion battery cathodes. Powder Technol 297:266–274. https://doi.org/10.1016/j.powtec.2016.04.011

Bonou A, Olsen SI, Hauschild MZ (2015) Introducing life cycle thinking in product development—a case from Siemens Wind Power. CIRP Ann Manuf Technol 64:45–48. https://doi.org/10.1016/j.cirp.2015.04.053

Bovea MD, Pérez-Belis V (2012) A taxonomy of ecodesign tools for integrating environmental requirements into the product design process. J Clean Prod 20:61–71. https://doi.org/10.1016/j.jclepro.2011.07.012

Broch F, Warsen J, Krinke S (2015) Implementing life cycle engineering in automotive development as a helpful management tool to support design for environment. Life cycle management. Springer International Publishing, Cham, pp 319–329

Cerdas F, Thiede S, Juraschek M et al (2017) Shop-floor life cycle assessment. In: Procedia CIRP, pp 393–398

Cerdas F, Andrew S, Thiede S, Herrmann C (2018a) Environmental aspects of the recycling of lithium-ion traction batteries. In: Lithorec, pp 267–288

Cerdas F, Egede P, Herrmann C (2018b) LCA of electromobility. Life Cycle Assess

Cerdas F, Egede P, Herrmann C (2018c) LCA of electromobility. Life Cycle Assess 669–693. https://doi.org/10.1007/978-3-319-56475-3_27

Cerdas F, Gerbers R, Andrew S et al (2018d) Disassembly planning and assessment of automation potentials for lithium-ion batteries. In: Recycling of lithium-ion batteries the LithoRec way, pp 83–97

Cha JM, Suh SH (2011) Developing a conceptual framework for UT based LCA. In: Glocalized solutions for sustainability in manufacturing—Proceedings of the 18th CIRP international conference on life cycle engineering, pp 587–588

Chan CC, Chau KT (1997) An overview of power electron in electric vehicles. IEEE Trans Ind Electron 44:3–13. https://doi.org/10.1109/41.557493

Chertow MR (2000) The IPAT equation and its variants. J Ind Ecol 4:13–29. https://doi.org/10.1162/108819800525419277

CIRP (2017) STC A—life cycle engineering and assembly. Int Acad Prod Eng. https://www.cirp.net/scientific-groups/stc-scientific-technical-commitees/stc-a-life-cycle-engineering-and-assembly.html. Accessed 19 Nov 2018

Cox B, Mutel CL, Bauer C et al (2018) Uncertain environmental footprint of current and future battery electric vehicles. Environ Sci Technol 52:4989–4995. https://doi.org/10.1021/acs.est.8b00261

Crawley E, Cameron B, Selva D (2015) Systems architecture: strategy and product development for complex systems. Global edi. Pearson, Boston

Crutzen PJ (2002) Geology of mankind. Nature 415:23. https://doi.org/10.1038/415023a

Curran MA (2014) Strengths and limitations of life cycle assessment. In: Background and future prospects in life cycle assessment, pp 189–206

Daly HE (1990) Toward some operational principles of sustainable development. Ecol Econ 2:1–6. https://doi.org/10.1016/0921-8009(90)90010-R

Daly HE (1992) Allocation, distribution, and scale: towards an economics that is efficient, just, and sustainable. Ecol Econ 6:185–193. https://doi.org/10.1016/0921-8009(92)90024-M

de Haes HAU, Jolliet O, Finnveden G et al (1999) Best available practice regarding impact categories and category indicators in life cycle impact assessment. Int J Life Cycle Assess 4:66–74. https://doi.org/10.1007/BF02979403

Deng D (2015) Li-ion batteries: basics, progress, and challenges. Energy Sci Eng 3:385–418. https://doi.org/10.1007/978-94-007-1186-0_26

Dhillon BS (2009) Life cycle costing for engineers

Diekmann J, Hanisch C, Froböse L et al (2017) Ecological recycling of lithium-ion batteries from electric vehicles with focus on mechanical processes. J Electrochem Soc 164:A6184–A6191. https://doi.org/10.1149/2.0271701jes

Dorn R, Schwartz R, Steurich B (2013) Batteriemanagementsystem. Handbuch lithium-ionen-batterien. Springer, Berlin Heidelberg, Berlin, Heidelberg, pp 177–187

Duce AD, Egede P, Öhlschläger G et al (2013) Guidelines for the LCA of electric vehicles. European Union Seventh …, Brussels

Duflou J, Dewulf W, Sas P (2003) Pro-active life cycle engineering support tools. CIRP Ann Manuf Technol 52:29–32. https://doi.org/10.1016/S0007-8506(07)60523-X

EC-JRC (2010b) ILCD handbook: framework and requirements for life cycle impact assessment models and indicators

EC-JRC (2010a) International reference life cycle data system (ILCD). In: Handbook—general guide for life cycle assessment. Detailed guidance

EEA (2018) Electric vehicles from life cycle and circular economy perspectives—TERM 2018: Transport and Environment Reporting Mechanism (TERM) report

Egede P, Dettmer T, Herrmann C, Kara S (2015) Life cycle assessment of electric vehicles—a framework to consider influencing factors. Procedia CIRP 29:233–238. https://doi.org/10.1016/j.procir.2015.02.185

Egede P (2016) Environmental assessment of lightweight electric vehicles, pp 94–98

Egede P (2017) Electric vehicles, lightweight design and environmental impacts. Springer, pp 9–40

Ellingsen LAW, Hung CR, Majeau-Bettez G et al (2016) Nanotechnology for environmentally sustainable electromobility. Nat Nanotechnol 11:1039–1051. https://doi.org/10.1038/nnano.2016.237

Emilsson E, Dahllöf L (2019) Lithium-ion vehicle battery production, C444. IVL Swedish Environmental Research Institute

Gallagher KG, Nelson P (2014) Manufacturing costs of batteries for electric vehicles. Elsevier

Goodenough JB (1998) General concepts. Lithium ion batteries fundamentals and performance. Wiley-VCH Verlag GmbH & Co. KGaA, Weinheim, Germany, pp 1–25

Goodland R (1995) The concept of environmental sustainability. Annu Rev Ecol Syst 26:1–24. https://doi.org/10.1038/sj.bjc.6605745

Goodland R, Bank W (2002) Sustainability: human, social, economic and environmental. Soc Sci 6:220–225

Granta Design Limited (2013) CES selector software

Gregory JR, Noshadravan A, Olivetti EA, Kirchain RE (2016) A methodology for robust comparative life cycle assessments incorporating uncertainty. Environ Sci Technol 50:6397–6405. https://doi.org/10.1021/acs.est.5b04969

Groen EA, Heijungs R, Bokkers EAM, de Boer IJM (2014) Methods for uncertainty propagation in life cycle assessment. Environ Model Softw 62:316–325. https://doi.org/10.1016/j.envsoft.2014.10.006

Groen EA, Bokkers EAM, Heijungs R, de Boer IJM (2017) Methods for global sensitivity analysis in life cycle assessment. Int J Life Cycle Assess 22:1125–1137. https://doi.org/10.1007/s11367-016-1217-3

Grunditz E, Thiringer T (2016) Performance analysis of current BEVs—based on a comprehensive review of specifications. IEEE Trans Transp Electrif 7782:1–1. https://doi.org/10.1109/TTE.2016.2571783

Gutowski TG (2011) Manufacturing and the science of sustainability. In: Glocalized Solutions Sustainability in Manufacturing: Proceeding of the 18th CIRP International Conference Life Cycle Engineering, pp 32–39. https://doi.org/10.1007/978-3-642-19692-8-6

Guzzella L, Sciarretta A (2013) Vehicle propulsion systems. Springer, Berlin Heidelberg, Berlin, Heidelberg

Hambitzer G, Pinkwart K, Ripp C, Schiller C (2007) Thermodynamics and mechanistics. Handbook of battery materials. Wiley-VCH Verlag GmbH, Weinheim, Germany, pp 1–17

Hanisch C, Diekmann J, Stieger A et al (2015) Recycling of lithium-ion batteries. In: Handbook of Clean Energy System. https://doi.org/10.1002/9781118991978.hces221

Hauschild MZ (2015) Better—But is it good enough? On the need to consider both eco-efficiency and eco-effectiveness to gauge industrial sustainability. Procedia CIRP 29:1–7. https://doi.org/10.1016/j.procir.2015.02.126

Hauschild M, Jeswiet J, Alting L (2005) From life cycle assessment to sustainable production: status and perspectives. CIRP Ann Manuf Technol 54:1–21. https://doi.org/10.1016/S0007-8506(07)60017-1

Hauschild MZ, Huijbregts MAJ (2015) Introducing life cycle impact assessment. In: Life cycle impact assessment, pp 1–16

Hauschild MZ, Herrmann C, Kara S (2017a) An integrated framework for life cycle engineering. Procedia CIRP 61:2–9. https://doi.org/10.1016/j.procir.2016.11.257

Hauschild MZ, Rosenbaum RK, Olsen SI (2017b) Life cycle assessment: theory and practice. ieeexplore.ieee.org

Hauschild MZ (2018) Introduction to LCA. In: Life Cycle Assessment, pp 59–66

Hauschild MZ, Bonou A, Olsen SI (2018) Life Cycle interpretation. Life cycle assessment. Springer International Publishing, Cham, pp 323–334

Hawkins TR, Singh B, Majeau-Bettez G, Strømman AH (2013) Comparative environmental life cycle assessment of conventional and electric vehicles. J Ind Ecol 17:53–64. https://doi.org/10.1111/j.1530-9290.2012.00532.x

Heijungs R, Lenzen M (2014) Error propagation methods for LCA—A comparison. Int J Life Cycle Assess 19:1445–1461. https://doi.org/10.1007/s11367-014-0751-0

Heijungs R, Suh S (2002) The Computational structure of life cycle assessment. Springer, Netherlands, Dordrecht

Hellweg S, Mila i Canals L (2014) Emerging approaches, challenges and opportunities in life cycle assessment. Science (80-) 344:1109–1113. https://doi.org/10.1126/science.1248361

Herrmann C, Hauschild M, Gutowski T, Lifset R (2014) Life cycle engineering and sustainable manufacturing. J Ind Ecol 18:471–477. https://doi.org/10.1111/jiec.12177

Herrmann F, Rothfuss F (2015) Introduction to hybrid electric vehicles, battery electric vehicles, and off-road electric vehicles. Elsevier Ltd.

Herrmann C (2003) Unterstützung der Entwicklung recyclinggerechter Produkte. Technische Universität Braunschweig

Herrmann C (2010) Ganzheitliches life cycle management

Hofer J (2014) Sustainability assessment of passenger vehicles: analysis of past trends and future impacts of electric powertrains. research-collection.ethz.ch

Hofmann P (2010) Hybridfahrzeuge. Springer Vienna, Vienna

Holdren JP, Ehrlich PR (1971) Impact of population growth. Am Assoc Adv Sci 171:1212–1217. https://doi.org/10.1017/CBO9781107415324.004

Huijbregts MAJ, Gilijamse W, Ragas AMJ, Reijnders L (2003) Evaluating uncertainty in environmental life-cycle assessment. A case study comparing two insulation options for a Dutch one-family dwelling. Environ Sci Technol 37:2600–2608. https://doi.org/10.1021/es020971+

European Commission—Joint Research Centre—Institute for Environment and Sustainability (2010) ILCD handbook: general guide for life cycle assessment—Detailed guidance

Ishii K (1995) Life-cycle engineering design. J Mech Des 117:42. https://doi.org/10.1115/1.2836469

Islam R, Husain I, Fardoun A, McLaughlin K (2009) Permanent-magnet synchronous motor magnet designs with skewing for torque ripple and cogging torque reduction. IEEE Trans Ind Appl 45:152–160. https://doi.org/10.1109/TIA.2008.2009653

ISO (2006) ISO 14040-environmental management—life cycle assessment—principles and framework. Int Organ Stand 3:20. https://doi.org/10.1016/j.ecolind.2011.01.007

IUCN (2008) the IUCN Programme 2005—2008 many voices, one earth

Jeswiet J, Hauschild M (2005) Eco-Design and future environmental impacts. Mater Des 26:629–634. https://doi.org/10.1016/j.matdes.2004.08.016

Kaiser J, Wenzel V, Nirschl H et al (2014) Prozess- und Produktentwicklung von Elektroden für Li-Ionen-Zellen. Chem-Ing-Tech 86:695–706. https://doi.org/10.1002/cite.201300085

Kambly KR, Bradley TH (2014) Estimating the HVAC energy consumption of plug-in electric vehicles. J Power Sources 259:117–124. https://doi.org/10.1016/j.jpowsour.2014.02.033

Kampker A (2014) Elektromobilproduktion. Springer, Berlin Heidelberg, Berlin, Heidelberg

Kampker A, Hohenthanner C, Deutskens C et al (2013b) Handbuch lithium-ionen-batterien. Springer, Berlin Heidelberg, Berlin, Heidelberg

Kampker A, Vallée D, Schnettler A (2013a) Elektromobilität

Kara S (2009) Life cycle engineering: applying life cycle knowledge to engineering solutions. CIRP J Manuf Sci Technol 1:213. https://doi.org/10.1016/j.cirpj.2009.07.001

Kara S, Hauschild MZ, Herrmann C (2018) Target-driven life cycle engineering: staying within the planetary boundaries. Procedia CIRP 69:3–10. https://doi.org/10.1016/j.procir.2017.11.142

Karden E, Ploumen S, Fricke B et al (2007) Energy storage devices for future hybrid electric vehicles. J Power Sources 168:2–11. https://doi.org/10.1016/j.jpowsour.2006.10.090

Keiner M (2006) The future of sustainability. Futur Sustain 1–257. https://doi.org/10.1007/1-4020-4908-0

Keoleian GA (1993) The application of life cycle assessment to design. J Clean Prod 1:143–149. https://doi.org/10.1016/0959-6526(93)90004-U

Ketterer B, Karl U, Möst D, Ulrich S (2009) Lithium-ionen batterien: Stand der Technik und Anwendungspotenzial in Hybrid-, Plug-In Hybrid-und Elektrofahrzeugen. Wissenschaftliche Berichte FZKA 7503:118

Kim SJ, Kara S (2014) Predicting the total environmental impact of product technologies. CIRP Ann Manuf Technol 63:25–28. https://doi.org/10.1016/j.cirp.2014.03.007

Kim SJ, Kara S, Hauschild M (2017) Functional unit and product functionality—addressing increase in consumption and demand for functionality in sustainability assessment with LCA. Int J Life Cycle Assess 22:1257–1265. https://doi.org/10.1007/s11367-016-1233-3

Klöpfer W, Grahl B (2014) Life cycle assessment (LCA) a guide to best practice. Wiley-VCH Verlag Gmbh & Co, KGaA, Weinheim, Germany

Klöpffer W, Grahl B (2014) Life cycle assessment (LCA). Wiley-VCH Verlag GmbH & Co, KGaA, Weinheim, Germany

Korthauer R (2013) Handbuch lithium-ionen-batterien. Springer, Berlin Heidelberg, Berlin, Heidelberg

Kwade A, Haselrieder W, Leithoff R et al (2018) Current status and challenges for automotive battery production technologies. Nat Energy 3:290–300. https://doi.org/10.1038/s41560-018-0130-3

Lacirignola M, Blanc P, Girard R et al (2017) LCA of emerging technologies: addressing high uncertainty on inputs' variability when performing global sensitivity analysis. Sci Total Environ 578:268–280. https://doi.org/10.1016/j.scitotenv.2016.10.066

Laurin L, Amor B, Bachmann TM et al (2016) Life cycle assessment capacity roadmap (section 1): decision-making support using LCA. Int J Life Cycle Assess 21:443–447. https://doi.org/10.1007/s11367-016-1031-y

Lesage P, Muller S (2017) Life cycle inventory: an in-depth look at the modeling, data, and available tools. In: Encyclopedia of sustainable technologies. Elsevier, pp 267–275

Lesage P, Mutel C, Schenker U, Margni M (2018) Uncertainty analysis in LCA using precalculated aggregated datasets. Int J Life Cycle Assess 1–18. https://doi.org/10.1007/s11367-018-1444-x

Lindahl M, Ekermann S (2013) Structure for categorization of eco-design methods and tools. Re-engineering Manuf Sustain 117–122. https://doi.org/10.1007/978-981-4451-48-2_19

Liu W (2017) Hybrid electric vehicle system modeling and control, 2nd edn. Wiley, Chichester, West Sussex, UK

Mansour M (2006) Unterstützung der lebenszyklusorientierten Produktentwicklung durch Informations- und Wissensbereitstellung. Technischen Universität Carolo-Wilhelmina zu Braunschweig

Meyer C, Bockholt H, Haselrieder W, Kwade A (2017) Characterization of the calendering process for compaction of electrodes for lithium-ion batteries. J Mater Process Technol 249:172–178. https://doi.org/10.1016/j.jmatprotec.2017.05.031

Mi C, Masrur MA, Gao DW (2011) Hybrid electric vehicles: principles and applications with practical perspectives, 2nd edn. Wiley, Hoboken, NJ, USA

Michaelis S, Rahimzei E, Kampker A et al (2018) Roadmap Batterie-Produktionsmittel 2030. VDMA

Moltesen A, Bjørn A (2018) LCA and sustainability. Life Cycle Assessment. Springer International Publishing, Cham, pp 43–55

Nishio K, Furukawa N (2007) Practical batteries. In: Handbook of battery materials. pp 19–61

Nitta N, Yushin G (2014) High-capacity anode materials for lithium-ion batteries: choice of elements and structures for active particles. Part Part Syst Charact 31:317–336. https://doi.org/10.1002/ppsc.201300231

Nitta N, Wu F, Lee JT, Yushin G (2014) Li-ion battery materials: present and future. Mater Today 18:252–264. https://doi.org/10.1016/j.mattod.2014.10.040

Nordelöf A, Messagie M, Tillman AM et al (2014) Environmental impacts of hybrid, plug-in hybrid, and battery electric vehicles—what can we learn from life cycle assessment? Int J Life Cycle Assess 19:1866–1890

Nordelöf A, Tillman AM (2018) A scalable life cycle inventory of an electrical automotive traction machine—Part II: manufacturing processes. Int J Life Cycle Assess

Nordelöf A, Tillman AM, Grunditz E et al (2018) A scalable life cycle inventory of an electrical automotive traction machine—Part II: manufacturing processes. Int J Life Cycle Assess 23:295–313. https://doi.org/10.1007/s11367-017-1309-8

Owsianiak M, Bjørn A, Laurent A et al (2018) LCA applications. Life Cycle Assessment. Springer International Publishing, Cham, pp 31–41

Parliamentary Commissioner for the Environment (2002) Creating our future, sustainable development for New Zealand

Pearce DW, Atkinson GD (1993) Capital theory and the measurement of sustainable development: an indicator of "weak" sustainability. Ecol Econ 8:103–108. https://doi.org/10.1016/0921-8009(93)90039-9

Pearce DW, Atkinson GD (2010) The Concept of sustainable development: an evaluation of its usefulness ten years after Brundtland. CSERGE Work Pap PA 98–02:30–32. https://doi.org/10.1086/381115

Placke T, Kloepsch R, Dühnen S, Winter M (2017) Lithium ion, lithium metal, and alternative rechargeable battery technologies: the odyssey for high energy density. J Solid State Electrochem 21:1939–1964. https://doi.org/10.1007/s10008-017-3610-7

Rahimzei E, Sann K, Vogel M (2015) Kompendium : Li-ionen-Batterien Grundlagen, Bewertungskriterien, gesetze und normen

Rebitzer G, Ekvall T, Frischknecht R et al (2004) Life cycle assessment Part 1: Framework, goal and scope definition, inventory analysis, and applications. Environ Int 30:701–720. https://doi.org/10.1016/j.envint.2003.11.005

Rockström J (2015) Bounding the planetary future: why we need a great transition. Gt Transit Initiat 1–13

Rosenbaum RK, Georgiadis S, Fantke P (2018) Uncertainty management and sensitivity analysis. Life cycle assessment. Springer International Publishing, Cham, pp 271–321

Rossi M, Germani M, Zamagni A (2016) Review of ecodesign methods and tools. Barriers and strategies for an effective implementation in industrial companies. J Clean Prod 129:361–373. https://doi.org/10.1016/j.jclepro.2016.04.051

Rousseaux P, Gremy-Gros C, Bonnin M et al (2017) "Eco-tool-seeker" A new and unique business guide for choosing eco-design tools. J Clean Prod 151:546–577. https://doi.org/10.1016/j.jclepro.2017.03.089

Sacchi R, Besseau R, Perez-Lopez P, Blanc I (2019) Exploring technologically, temporally and geographically-sensitive life cycle inventories for wind turbines: a parameterized model for Denmark. Renew Energy

Schilling A, Schmitt J, Dietrich F, Dröder K (2016) Analyzing bending stresses on lithium-ion battery cathodes induced by the assembly process. Energy Technol 4:1502–1508. https://doi.org/10.1002/ente.201600131

Schmitt J, Raatz A, Dietrich F et al (2014) Process and performance optimization by selective assembly of battery electrodes. CIRP Ann 63:9–12. https://doi.org/10.1016/j.cirp.2014.03.018

Schmuch R, Wagner R, Hörpel G et al (2018) Performance and cost of materials for lithium-based rechargeable automotive batteries. Nat Energy 3:267–278. https://doi.org/10.1038/s41560-018-0107-2

Schönemann M (2017a) Multiscale simulation approach for battery production systems

Schönemann M (2017b) Multiscale simulation approach for battery production systems

Telenko C, Seepersad CC, Webber ME (2008) A compilation of design for environment principles and guidelines. In: Volume 5: 13th design for manufacturability and the lifecycle conference; 5th symposium on international design and design education; 10th international conference on advanced vehicle and tire technologies. ASME, pp 289–301

Tschöke H (2015) Die Elektrifizierung des Antriebsstrangs. Springer Fachmedien Wiesbaden, Wiesbaden

United Nations (2002) The Johannesburg declaration on sustainable development. Johaneesbg plan Implement, pp 1–62

Un-noor F, Padmanaban S, Mihet-popa L et al (2017) A comprehensive study of key electric vehicle (EV) components, technologies, challenges, impacts, and future direction of development. Energies 10:1–82. https://doi.org/10.3390/en10081217

Van Loon P, Olsson L, Klintbom P (2018) LCA guidelines for electric vehicles. Swedish Energy Agency, pp 1–32

VDMA W (2012) Montageprozess eines Batteriepacks. Der Montageprozess eines Batter Werkzeugmaschinenlabor WZL der RWTH Aachen und Verband Dtsch Maschinen- und Anlagenbau, 1 Aufl WZL und VDMA Eig Aachen

Velázquez-Martínez O, Valio J, Santasalo-Aarnio A et al (2019) A critical review of lithium-ion battery recycling processes from a circular economy perspective. Batteries 5:5–7. https://doi.org/10.3390/batteries5040068

Verones F, Hellweg S, Hauschild M et al (2015) LC-impact version 0.5

Wang H, Friedrich B (2015) Development of a highly efficient hydrometallurgical recycling process for automotive Li–ion batteries. J Sustain Metall 1:168–178. https://doi.org/10.1007/s40831-015-0016-6

Wanyama W, Ertas A, Zhang HC, Ekwaro-Osire S (2003) Life-cycle engineering: Issues, tools and research. Int J Comput Integr Manuf 16:307–316. https://doi.org/10.1080/0951192031000089255

Wegener K, Andrew S, Raatz A et al (2014) Disassembly of electric vehicle batteries using the example of the Audi Q5 hybrid system. Procedia CIRP 23:155–160. https://doi.org/10.1016/j.procir.2014.10.098

Wei W, Larrey-Lassalle P, Faure T et al (2015) How to conduct a proper sensitivity analysis in life cycle assessment: taking into account correlations within LCI data and interactions within the LCA calculation model. Environ Sci Technol 49:377–385. https://doi.org/10.1021/es502128k

Weidema BP (2001) Areas of protection and the impact chain. https://lca-net.com/files/impact-chain.pdf, 1–6

Westkamper E, Alting L, Arndt G (2000) Life cycle management and assessment: approaches and visions towards sustainable manufacturing. CIRP Ann Manuf Technol 49:501–522

Westkämper E, Alting A (2000) Life cycle management and assessment: approaches and visions towards sustainable manufacturing (keynote paper). CIRP Ann Manuf Technol 49:501–526. https://doi.org/10.1016/S0007-8506(07)63453-2

Woodward DG (1997) Life cycle costing—theory information acquisition and application. Int J Project Mange 15:335–344. https://doi.org/10.1016/S0263-7863(96)00089-0

World Commission on Environment and Development (1987) Our common future

Zeyen MG, Wiebelt A (2013) Thermisches Management der Batterie. Handbuch lithium-ionen-batterien. Springer, Berlin Heidelberg, Berlin, Heidelberg, pp 165–175

Chapter 3
State of Research—Review on LCE Modelling and Assessment Approaches for Electromobility

Contents

As seen in Chap. 2, a comprehensive evaluation and analysis of the environmental impacts of EVs and traction batteries is highly time intensive. An overarching programmatic approach to enable the fast and reliable life cycle-oriented tailoring and integration of interdisciplinary models is currently absent. To understand the potential of such an approach, it is essential to consult the approaches and technical developments recently documented within the research community. The aim of this chapter is, therefore, to review the contemporary research most relevant to the topic in question. The outline is as follows. First, the criteria to select and evaluate the approaches is presented. Then, the selected contributions are described and briefly analyzed. A comparative evaluation follows and, finally, the research gap is identified and described.

F. Cerdas, *Integrated Computational Life Cycle Engineering for Traction Batteries*, Sustainable Production, Life Cycle Engineering and Management,
https://doi.org/10.1007/978-3-030-82934-6_3

3.1 Selection of Approaches and Definition of Evaluation Criteria

Research on the assessment of potential environmental implications of electromobility has seen an important increase within the last few years (Hawkins et al. 2012; Larcher and Tarascon 2014; Nordelöf et al. 2014; Nealer and Hendrickson 2015; Ellingsen et al. 2017; Peters et al. 2017). However, although modelling approaches, as well as the computational tools and frameworks supporting these evaluations, have emerged, they have not been sufficiently developed to properly address all of the relevant research questions of the field. On the other hand, there have been several general developments concerning the computational modelling of complex systems which have the potential to improve the way LCE modelling is executed. Accordingly, the literature analysis presented in this chapter aims at giving an overview to the following questions: (1) what are the main computational tools and modelling approaches applied to and/or developed to generate knowledge on the environmental impact of electromobility? (2) what is the state of knowledge regarding the environmental impacts of electric vehicles and traction batteries? And (3) which current general modelling technologies within the discipline have the potential to improve computational modelling in LCE for electromobility?

3.1.1 *Selection of Approaches*

Derived from these three questions, a categorization framework was defined (Fig. 3.1). The scope of the literature review has been framed within a wider spectrum of disciplines including *sustainability assessment, industrial ecology, LCA, LCE and eco-design.* In this regard, the main branch of knowledge within which this research is framed is the field of *environmental assessment of electric vehicles and traction*

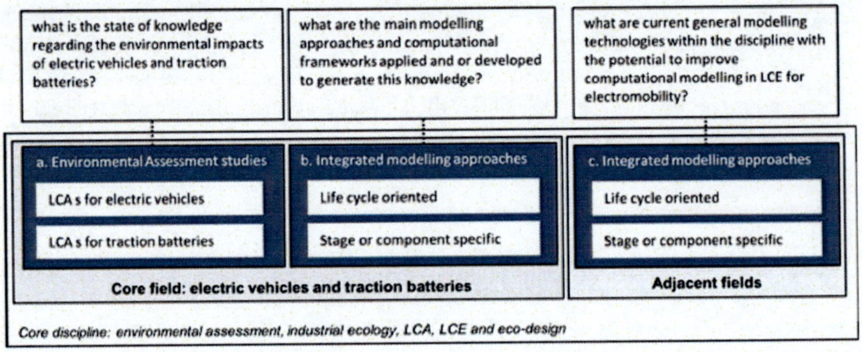

Fig. 3.1 Categorization framework of scientific literature and technological developments reviewed

batteries. Adjacent fields within the discipline were additionally explored with the objective of finding relevant computational modelling approaches (see Fig. 3.1).

As seen in Fig. 3.1, three main categories were defined as relevant to the overall aim of this thesis:

a. *Environmental Assessment studies*: this group refers to studies that exclusively aim at estimating the potential environmental implications rising from electromobility. Specifically, the focus of this category is the analysis of assessment studies evaluating the impacts of the production of traction batteries and electric vehicles from a life cycle perspective.
b. *Modeling approaches within the core field*: in this group, all relevant modelling approaches (i.e. tools and software) that are applied (or were developed for that purpose) within the assessment studies will be investigated. The focus here lies not only on models intended to specifically estimate the environmental impacts of EVs or traction batteries, but also on models depicting particular stages within their life cycle (e.g. manufacturing of cells, production of cathode materials, use phase, among others). While the objective of some of these models are not to support environmental assessment per se, many provide useful information regarding, for instance, material and energy flows.
c. *Modelling approaches in adjacent fields*: Looking beyond the nature of the case study, the intention of this category is to review modelling approaches in adjacent fields, with the objective of finding similarities or lessons learned which might contribute to the development of the concept in this work.

Current literature in the field of the environmental issues of electromobility is both vast and diverse. For the categories (a) and (b) a first screening[1] of the literature resulted in a total of 224 contributions. A social network analysis (SNA)[2] was further

[1] A structured literature search was performed using the querying service from google scholar. For the first category, *environmental assessment studies,* the queries were done by searching for documents including all of the terms "Life Cycle Assessment", "LCA", "Environmental impacts", "emissions", "transportation" and any of the terms "industrial ecology", "eco-design", "environmental assessment", "Life Cycle Engineering" and "LCE" within the document's keywords. In this regard, one query was done including the term "electric vehicle" in the document's title and a second one including the term "battery". This resulted in a total of 362 and 220 contributions respectively. For the second category, *modeling approaches in the field of LCA for electromobility* the list of keywords was complemented with the terms "electric vehicle" and "battery". One query was done including the term "model" in the document's title and a second one including the term "approach". This resulted in a total of 120 and 113 contributions respectively. In total, four different queries were executed resulting in 815 results. These were systematically filtered according to the following exclusion criteria: (1) older than 2011, (2) strict focus on the assessment of technical performance, (3) focus on business or strategy (e.g. market penetration studies, policy analysis, strategic planning), (4) focus on operations research and optimization, (5) focus on cost analysis, (6) focus on grid integration and renewable energy focus, (7) conference papers, (8) repeated query results and (9) papers with no citations. Further, the list was complemented with contributions subjectively selected from three of the excluded groups: (1), (5) and (9). Finally, the list of relevant was pre-filtered to a total 224 contributions.

[2] SNA is an approach to research social structures by applying concepts from network and graph theory. It provides a structured method to describe networks as a set of nodes (main actors or

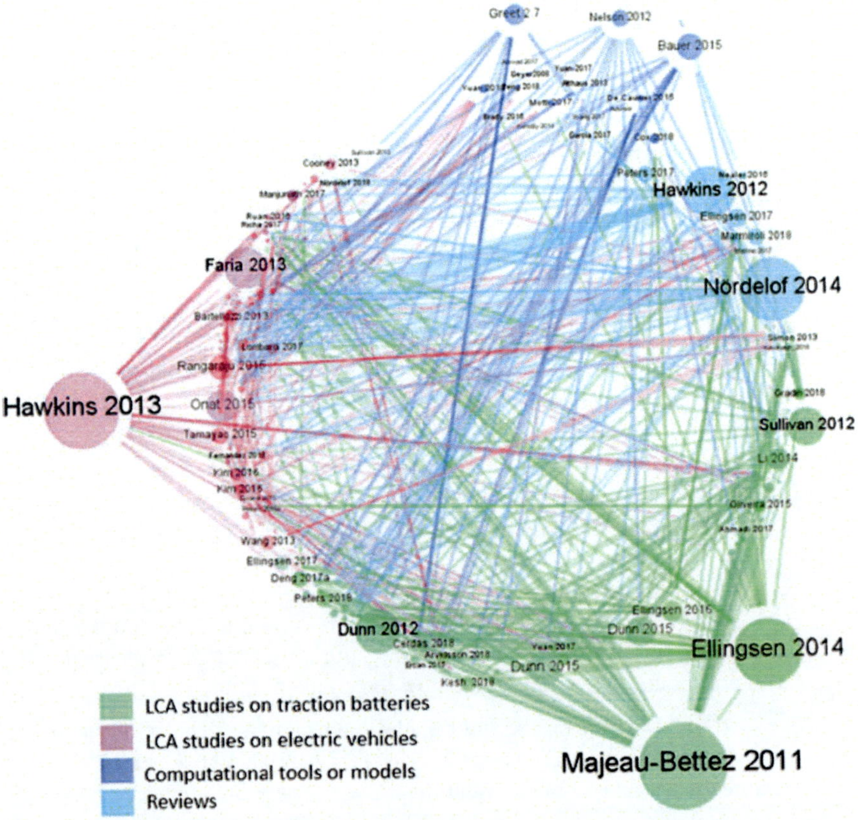

Fig. 3.2 Visual analysis of the citation network structure within the field of Environmental Assessment of EV and traction batteries. The size of the node represents the number of citations a paper has within the network. The color of the edge matches the source node's color

performed, with the goal of identifying key players within the field. A visual representation of the citation network is shown in Fig. 3.2. All nodes were programmatically clustered as LCA studies on batteries (green nodes), LCA studies on EVs (magenta nodes), studies which described the development of a software or computational tool to perform such assessments (blue nodes) and review studies (cyan nodes). The graph shows the most influential[3] publications within the set of contributions previously filtered. The analysis served as a basis to further filter approaches that, though they

individuals) and edges (interactions between the nodes) (Otte and Rousseau 2002). In this case, each of the contributions identified in the structure literature search was assigned a node. The nodes are further linked through edges to the contributions citing or making use of the information in the paper.

[3] Influence was measured as eigenvector centrality. In graph theory, a node with a high eigenvector score implies that the node is linked to a large number of nodes which have, in turn a high eigenvector centrality.

differ from one another in terms of assessment scope, do not contribute to the discussion required in this thesis. Using these filters, a selection of high-influence and low influence contributions was further created for the first two categories (a and b).

Highly cited studies

Regarding modelling approaches and/or computational frameworks, specific tools such as the Greenhouse Gases, Regulated Emissions, and Energy Use in Transportation (GREET Model) (Burnham et al. 2006) and the Battery Performance and Cost Model (BatPac Model) (Nelson et al. 2011), both from the Argonne National Laboratory, were found to have an important influence within the scientific literature in the field of LCA for electromobility. The influence and relevance of contributions such as the ones from Hawkins and colleagues (2013), Majeau-Bettez and colleagues (2011), Ellingsen et al. (2014), to mention a few, lies primarily on the transparency and comprehensiveness of the LCI presented within the publication. These publications, among others which included an open LCI, made possible the rise of many of other LCA studies in the field of electromobility.

Less cited studies

From the less influential group of publications, a further manual selection was done, paying particular attention to studies that focused on important research questions that were not specifically addressed by the highly cited studies. In the case of LCAs for EVs, additional research exploring the effect of local context on the life cycle environmental impacts of EVs were screened and selected. Also, papers addressing the influence of real driving patterns to the energy consumption during the usage stage of a vehicle were searched for. For the case of LCAs for traction batteries, additional studies were selected which focused on questions such as: the evaluation of further battery materials; a (more) detailed examination of materials supply chains; assessments of the cell manufacturing stage; analysis of battery degradation during the usage stage and its effect to the life cycle impacts; and finally the analysis of the significance of recycling. Additionally, integrated approaches developed with the objective of performing a specific assessment (as opposed to developing a tool per se) were also identified as having a relatively important influence. This is the case of the research done by Bauer and colleagues (2015) and, more recently, of Hofer (2014), Cox and colleagues (Cox and Mutel 2018; Cox et al. 2018).

For the third category (c), a structured literature search was performed as well. In this case, the search was restricted to papers published no earlier than 2017, with a focus on (environmental) sustainability assessment which presented a computational approach. The fields explored were mainly: chemical products, waste and water treatment, building sector, energy systems and renewable energy. A total of 13 approaches were identified and selected. All selected publications will be described with more detailed further in this chapter.

3.1.2 Derivation of Evaluation Criteria

This section introduces the evaluation criteria used to assess the selected approaches on their contribution to the state of research. The criteria were defined based on requirements that were identified as essential for LCE tools and methodologies (see Sect. 2.1.3) and the questions in which the categories were framed (Fig. 3.1). The contributions selected were evaluate according to the following criteria:

a. *General scope criteria*: these aim at classifying the main objective of the contribution. In this group three criterion were defined:

   - **Focus on environmental impacts of traction batteries**: *is the scope of the paper relevant to the field of electromobility?*
   - **Focus on life cycle oriented environmental impacts assessment**: *is the aim of the paper to evaluate the potential environmental impact of the product system from a life cycle perspective?*
   - **Focus on computational modelling:** *is the aim of the paper to develop a modelling approach or a computational framework for the assessment of environmental impacts?*
   - **Focus on the raw materials stage**: h*ow detailed is the consideration of the raw materials stage in the modelling approach*
   - **Focus on the manufacturing stage**: *how detailed is the consideration of the manufacturing stage in the modelling approach*
   - **Focus on the usage stage**: *how detailed is the consideration of the usage stage in the modelling approach*
   - **Focus on the end-of-life stage**: *how detailed is the consideration of the end-of-life stage in the modelling approach*
   - **Comprehensiveness of the LCI**: *how complete and transparent is the LCI presented in the paper?*

b. *Assessment-related criteria*: these aim at evaluating the extent of the assessment presented in the contributions. The focus lies specifically on the evaluation of the construction of the LCI presented. The following criterion were defined:

   - **Usability of the LCI**: *were models provided that enable to further use the LCI published?*
   - **Consideration of variability in the foreground system**: *does the study consider variability in the foreground system?*
   - **Consideration of variability in the background system**: *does the study consider variability in the background system?*
   - **Consideration of spatial variability**: *does the study consider spatial variability?*

c. *Modelling-related criteria*: these aim at evaluating the integration capability, and the extent and capability of the modelling approaches and computational frameworks available. The following criterion were defined:

- **Integration of informatics tasks**: *does the model support integration of informatics tasks?*
- **Visual and data analytics**: *does the model support integration of visual and data analytics routines?*
- **Statistical methods for the assessment of uncertainties and sensitivities**: *does the model support integration of advanced statistical methods for the assessment of uncertainties and sensitivities?*
- **Prospective modelling**: *does the model support scenario modelling (or the integration of models) for the assessment of future technologies?*
- **Scientific and empirical models**: *does the model support the integration of scientific and/or empirical foreground and background system models as well as models describing local contexts?*

3.2 Description and Evaluation of Selected Approaches

In this section, the approaches previously selected are described and discussed. The description of the contributions is presented, following the structure in Fig. 3.1. First, with a presentation of all selected modelling approaches and computational frameworks. Next, this is followed with a discussion of the most relevant assessment studies for traction batteries, and finally, an analysis of the relevant computational approaches in adjacent fields.

3.2.1 State of the Research on the Life Cycle Environmental Assessment of Electric Vehicles and Traction Batteries

The increased technological and economic relevance of EVs in the transportation market has given rise to many studies that aim to evaluate the potential environmental advantages and disadvantages that a widespread introduction of this technology could bring. Important reviews have been published in this regard, both on the application of LCA for the assessment of EVs at a vehicle level (Hawkins et al. 2012; Nordelöf et al. 2014; Nealer and Hendrickson 2015) and at battery level (Ellingsen et al. 2017; Peters et al. 2017; Peters and Weil 2018). Additionally, guidelines with the objective of providing standards for the definition of system boundaries, functional units and for the consideration of important influencing factors have been developed (Duce et al. 2013; Egede et al. 2015). The purpose of this section is not to provide a detailed discussion of the individual LCA studies and results. Rather, the focus of this analysis is to assess the scope of the contributions in terms of coverage of influencing factors in the foreground and background system; the usability of the inventory provided; the evaluation of the variability and sensitivity; and the general approach followed to model the product systems.

Table 3.1 Variability in the LCA results of EVs among the literature

Author	GWP [g CO_2-eq/km]	
	BEV	HEV
Aguirre et al. (2012)	110	145
Bauer et al. (2015)	60–375	240
Girardi et al. (2015)	155	–
Huo et al. (2015)	110–280	–
Bartolozzi et al. (2013)	110–175	–
Gao and Winfield (2012)	245	195
Tagliaferri et al. (2016)	145	150
Ellingsen et al. (2016)	250	–
Hawkins et al. (2013)	196–206	–
Faria et al. (2013)	75–315	–

LCA results on EVs vary largely within the literature available. As shown in Table 3.1, reported values range from approximately 60 g CO_2-eq/km to an upper boundary of about 400 g CO_2-eq/km.

This variation has been linked to differences regarding the modelling choices taken by the authors. Particularly decisive are: (i) factors regarding the scope of the study such as the definition of the functional unit, vehicle's lifetime and the system boundaries of the study, (ii) factors defining the foreground system such as the vehicle's weight, the size and configuration of the battery system (Ellingsen et al. 2016) and the approach used to estimate the energy required per kilometer (Li et al. 2016) and (iii) spatially differentiated factors such as weather and electricity mixes as summarized by Egede (Egede et al. 2015).

It is clear that LCA has been widely applied to evaluate the environmental impacts of EV.[4] The scope of these studies range from performing comparisons against conventional combustion engine vehicles (Hawkins et al. 2013) to evaluate the influence of the battery (Ellingsen et al. 2014; Cerdas et al. 2018b) or of a particular life cycle stage such as manufacturing (Thomitzek et al. 2019a, b) or recycling (Dunn et al. 2015a; Cerdas et al. 2018a). Although there are guidelines that structure the environmental assessment of such products, the complexity of a given modelling approach has led to simplifications that make the decision-making process difficult.

In the assessment of the environmental impacts of EVs, the contribution presented by Hawkins and colleagues (Hawkins et al. 2013) marked the beginning of a series of studies in which the assessment framework was increasingly defined beyond the vehicle's usage stage. Their study is, in this context, one of the flagship publications. In their study they provided transparently modelled life cycle inventories. The system boundary of the study was set as shown in Fig. 2.18, and the inventory was subsequently used to perform a life cycle assessment. In their study, they reported

[4] The most relevant research in this regard will be discussed in this chapter.

that EVs can potentially reduce GWP by around 20–24% when compared to gasoline ICEVs, and 10–14% when compared to diesel ICEVs, if the current European electricity mix is considered (Hawkins et al. 2013). They find the GWP impact created during the production stage of the electric vehicle is around twice as large as that of the conventional vehicle, with the production of the battery system represent around 40% (between 35 g CO_2-eq km^{-1} and 40 g CO_2-eq km^{-1}). While the results of the study enhanced the understanding of the environmental impacts of electric vehicles, the usability of their results and the capacity to make a more informed decision out of their results is limited. The authors found that the source of electricity used to power a EV has the largest GWP influence, causing the results to range from 106 g CO_2-eq km^{-1} for a vehicle powered only with wind power, to approximately 350 g CO_2-eq km^{-1} for an electricity mix predominantly from lignite combustion sources. They estimated the variability of the amount of fuel consumed per km by a conventional ICE vehicle causes the GWP results to vary from around 150 g CO_2-eq km^{-1} to around 320 g CO_2-eq km^{-1} for diesel and from 175 g CO_2-eq km^{-1} to 350 g CO_2-eq km^{-1} for gasoline. In the case of EVs, however, it was found that the variation of other parameters such as energy usage per km or "battery" also caused important variations of the GWP on a km basis. On the most important conclusions of the authors is that in current studies there is not enough data available to really understand the environmental impact of the battery system. In their research, the battery is understood as a parameter that can be varied in terms of mass and energy density. Their life cycle assessment models did not deeply explore their production processes or the relationships between product parameters, production parameters and environmental impacts. This was later explored by Ellingsen and colleagues (2014). They performed an LCA using primary data collected from the bill of materials taken from a pilot-line scale manufacturing plant of lithium-ion batteries, and combined that data with the life cycle inventory for battery cells presented by Majeau-Bettez et al. (2011), which was the same battery cell inventory used by Hawkins et al. (2013). They reported cradle to gate results for the battery pack ranging from 11 g CO_2-eq km^{-1} to 31 g CO_2-eq km^{-1}. However, as they reported, the results are strongly driven by the energy consumption (more than 60%) of the manufacturing process of the cells (Ellingsen et al. 2014). Additionally, other factors such as weather and driving profiles, which were subsequently identified by other researchers as important (Egede et al. 2015; Li et al. 2016) were not considered in this study.

As shown in Fig. 3.2, the contribution of Hawkins et al. (2013) has had a large influence upon ensuing LCA studies in the field. This is basically due two reasons: (i) It was one of the first studies to compare the environmental impacts of ICEVs and EVs while taking into consideration the whole life cycle of both vehicles and (ii) it provided a very comprehensive and transparent LCI. The scope of the study included the production, usage and end-of-life stages of a ICEV and an EV. The authors defined a generic vehicle glider for both technologies. Further, this glider was scaled to meet the mass and size of a Mercedes A-series for the ICEV and a Nissan Leaf for the EV. For the production stage, the authors make use of the software GREET™ (Burnham et al. 2006), along with some available specifications from the manufacturers of each model. Two types of battery systems were modelled in their study to match the power and energy requirements reported by the vehicle manufacturer, with the inventories for the NMC and LFP batteries adapted from the contribution from Majeau-Bettez et al. (2011). The usage stage was modelled considering only the energy consumption per km required for driving. In this regard, nominal information provided by the manufacturers was integrated into their inventory. As reported, the authors assumed a total of 0.173 kWh/km in the case of the EV, 68.5 ml/km of gasoline for the gasoline-powered ICEV and 53 ml/km of diesel for the diesel-powered ICEV. Both ICEV and EV were assumed to have a lifetime of 150,000 km disregarding the type of battery chemistry. Recycling was modelled using a dataset from battery treatment provided by the data provider ecoinvent (Weidema et al. 2013). Although the study presents a large number of rough assumptions, the authors were, nevertheless, able to derive important findings and identify necessary further research. As they reported, in the context of the European electricity mix, EVs can potentially reduce GWP by 10% to 24% relative to gasoline and diesel ICEV. Nevertheless, they found that EVs can be potentially more impactful in other categories such as human toxicity, eco-toxicity and eutrophication. Their results were found to be very sensitive to assumptions made regarding electricity mix, energy consumption per kilometer and vehicle lifetime, among others.

The research by Faria and colleagues (2013) takes a deeper look into the factors exerting a high influence on the environmental impacts of EVs within the usage stage. In their study, they real-world test compact (class B) and sub-compact (class A) vehicles. Using a data acquisition system which was installed in the vehicles, the authors were able to link specific user profiles to energy consumption patterns. Additionally, three different electricity mixes were included in the analysis: Polish (fossil fuel based), Portuguese (high content of hydropower, natural gas and wind) and French (mostly based on nuclear energy). Two important findings are discussed in their paper. First, the electricity mix with the highest degree of fossil-based energy (i.e. electricity mix from Poland) was found to have the largest influence on the EV's environmental impacts per kilometer. While this was expected, it was further found that both the ICEV model (diesel and gasoline) have a lower GWP than EVs connected to the polish electricity grid. Another important finding regards the effect of driving profiles. As reported in their research, an aggressive driving style can increase the energy consumed per kilometer by up to 47%. Moreover, the authors found that differences in the set-up of the cabin climate control implied variations of 20–60%.

Another study analyzing the way the electricity mix effects the environmental impact of EVs is presented by Rangaraju and colleagues (2015). They evaluated the effect of dynamic variations in electricity mixes and battery charging profiles on the WTT emissions of BEVs. In their study, the authors monitored the real-world driving data of five different BEVs within a period of two years, looking at energy consumption, charging patterns, and different driving styles. While the study is geographically restricted to the case of Belgium, the results are relevant to the general field of electromobility LCA, as they show that it is necessary to consider how the variability indifferent parameters of both background and foreground systems effects the environmental impacts of vehicles.

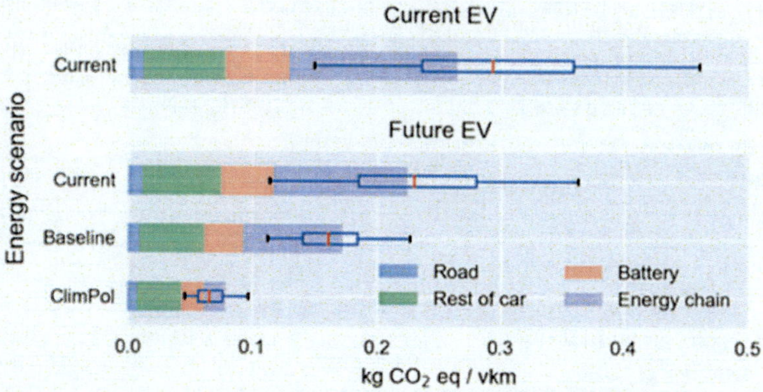

Fig. 3.3 Uncertain GWP of current and future BEV for several energy scenarios (Cox et al. 2018)

The contribution presented by Ellingsen and colleagues (2016) shows how variability in some of the most important influencing factors effects the foreground system. In their paper, the authors collected nominal information from manufacturers for four European segments (A, B, C, D, F) regarding curb weight, battery sizes, driving ranges and energy consumption per km, according the NEDC driving cycle. While the study is very simply in terms of data acquisition and the modelling approach followed, it shows that even small variations in terms of the range, size and weight of a vehicle can make a large difference in the cradle-to-grave life cycle emissions of EV.

While the work done by Cox and colleagues (2018) still did not considered the whole spectrum of influential factors, nor did it look at battery in detail, in terms of methodologically they made important progress towards putting together a model that could combine the combined effect of multiple factors on the life cycle GWP of an electric vehicle. Their main results are displayed in (Fig. 3.3).

It is fair to mention, though, that the objective of the paper was to evaluate the effect of future energy scenarios, and therefore important components like the battery were treated as in Ellingsen et al. (2014). Nevertheless, the authors presented a model that was able to estimate the combined effect of multiple influential parameters and, so, was able to estimate an even wider spectrum of possible GWP results, as shown in Fig. 2.23.

A constant in the literature regarding the environmental impacts of EVs is the lack of detail in which the battery system is modelled. Typically, the system is modelled by linking the mass of the battery required by the vehicle to the respective dataset provided by commercial life cycle inventory data bases such as ecoinvent (Steubing et al. 2016; Wernet et al. 2016), in which the battery cell consist of a LMO cathode. Aiming at increasing the resolution of the available LCA models for electric vehicles, several contributions have been published presenting cradle-to-gate models for the production of an EV battery system. As seen in Table 3.2, looking more closely at the battery system made it evident that the variability of the modelling approaches

Table 3.2 Variability in the LCA results of traction batteries among the literature

	kg CO2-eg/kWh Cell chemistry							
Author	(a)	(b)	(c)	(d)	(e)	(f)	(g)	(h)
Kim et al. (2016)	140							
Cusenza et al. (2019)	313							
Ellingsen et al. (2014)		172–487						
Majeau-Bettez et al. (2011)		200		250				
Li et al. (2014)		135	355					
Zackrisson et al. (2010)				166–266				
Notter et al. (2010)							52.6	
Dunn et al. (2012)		50					39	
Amarakoon et al. (2013)		120		150			60	
Faria et al. (2014)							70.9	
Cerdas et al. (2018b)		110						117
Deng et al. (2017)								140

(a) LMONCM/G, (b) NCM/G, (c) NCM/SiNW, (d) LFP/G, (e) LFP/LTO, (f) NCA/G, (g) LMO/G und (h) Li–S

and, consequently, the LCA, meant that results became higher and more complex to model. As seen, results ranging from 39 kg CO_2-eq up to around 480 kg CO_2-eq per kWh of battery capacity have been reported. This large difference in results reflects the difference of configurations in the of pairs of electrodes and materials, and the assumptions made to model or to quantify the energy consumption during the manufacturing stage which, as seen in Table 3.3, varies from 0.861 kWh to around 462 kWh.

Here, "top-down" refers to the quantification approach in which the total amount of energy demanded in a period of time is divided by the total energy storage capacity manufactured during this year. A "bottom-up" approach refers to the quantification

Table 3.3 Variability in the reported energy required during the manufacturing phase of battery cells

Author	kWh/kWh_{batt}	Modelling approach
Ellingsen et al. (2014)	162.7	Top-down
Notter et al. (2010)	0.861	Bottom-up
Zackrisson et al. (2010)	125.3	Top-down
Majeau-Bettez et al. (2011)	131.4	Top-down
Dunn et al. (2012)	2.97	Top-down
Yuan et al. (2017)	461.98	Bottom-up

of energy demand by summing up the individual demand of all the processes required to produce a given storage capacity.

The assumptions made to evaluate the material distribution at cell and battery pack level (see Table 3.4) has been very similar among the studies presenting the most transparent inventories. That said, important differences were found regarding, for example, the amount of material considered for the cell housing and other important components, such as the cooling system. Also worth noticing is the fact that there is a low granularity in the data regarding material configuration. For instance, there is no clear description of the amount of active material per unit of mass for cathode and anode paste, or the amount of solvents and binders.

The recycling stage is largely ignored by the LCA studies and approaches analyzed here. As with Nordelöf and colleagues (2019), throughout the scientific literature available, two main modelling approaches are followed to quantify the environmental

Table 3.4 Estimated values on material composition of battery systems among the literature

Battery components	A	B	C	D	E
	wt%	wt%	wt%	wt%	wt%
Cathode paste	23.2	–	22.8	–	25.0%
Cathode active material					
Anode paste	9.4		9.9		11.0%
Graphite					
Cell electrodes/collectors				40.0%	
Separator	3.3	2–3	1.3		3.0
Substrate cathode (Al)	3.6	4–9	2.9%		7.0
Substrate anode (Cu)	8.3	1–12	13.1%		8.0
Electrolyte	12.0	8–15	9.5		8.0
Cell electrolyte/separator		3–20		12.0	
Cell container	20.1		0.4		1.0
Steel tray/covers				22.0	
Steel panels/brackets				8.0	
Frame/brackets				7.0	
Module + batt. packaging	17.0	17–23			
Battery packaging			32.1		33.8
Enclosure					
BMS	3.0	1–2	3.7	3.0	3.2
Cooling system			4.1	3.0	
Electrical system				1.0	
Cell pouch/other				3.0	
Other battery components				0.5	

A: (Majeau-Bettez et al. 2011), B: (Amarakoon et al. 2013), C: (Ellingsen et al. 2014), D: (Kim et al. 2016), E: (Cerdas et al. 2018a, b)

relevance of recycling: (i) the cutoff approach, in which no credit is estimated in the modelling of the E-o-L stage. Here the benefits of recycling are measured through the recycled content built into the battery cell, (ii) the E-o-L recycling approach, which measures the benefits of recycling by quantifying the burden it avoids (Nordelöf et al. 2019). As an example, in a national flagship project called LithoRec (Kwade et al. 2016), the environmental assessment of the E-o-L stage was quantified following the avoided burden approach. In this project, the assessment was performed by analyzing real data from disassembly experiments, and from a pilot process chain composed of disassembly steps, mechanical treatment and hydrometallurgical steps. As shown in Fig. 3.4, the implications of a sound recycling process might affect the total life cycle impact of a traction battery system. In this case, while the recycling process itself implies an impact of over 2 tons of CO_2-eq to recover the materials of a 320 kg battery pack, the GWP avoided impact linked to the materials recovered is more than 4.5 tons of CO_2-eq.

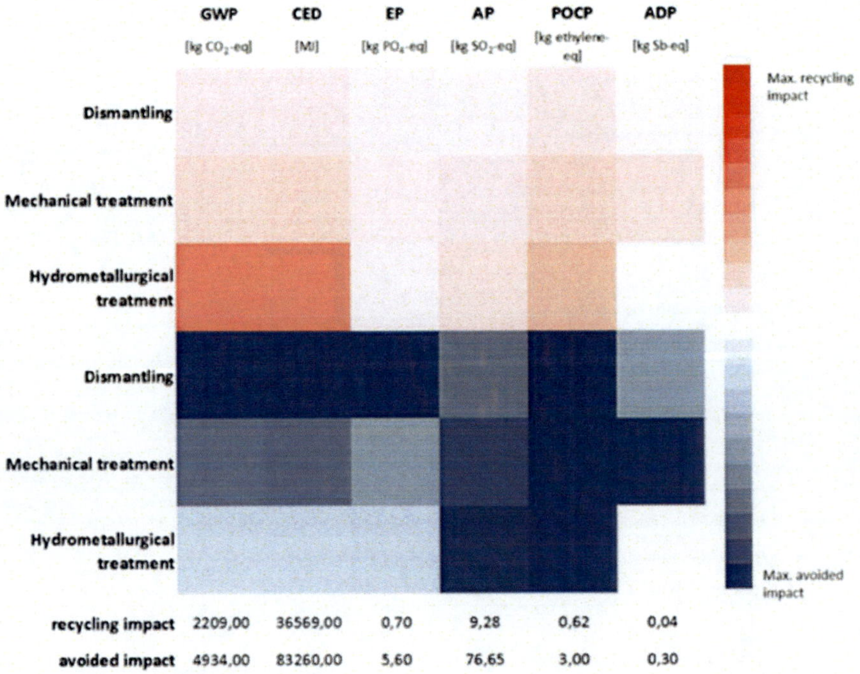

Fig. 3.4 Environmental implications of recycling in the project Lithorec (Cerdas et al. 2018a)

3.2.2 State of the Research on Selected Modelling Approaches and Computational Frameworks for EVs and Traction Batteries

In the last ten years, several modelling approaches partially implemented in software applications for the assessment of electromobility and traction batteries have emerged. In this literature analysis, the selected approaches are classified into environment-related, cost-related and performance-related models. These approaches focused on one or more life cycle stages. Software and tools focused on the assessment of cost and technical performance were additionally selected and analyzed based on their ability to estimate environmental impacts.

To start, a selected group of integrated assessment frameworks for the assessment of the whole vehicle is presented. The software ADVISOR™ was developed by the National Renewable Energy Laboratory NREL[5] with the objective of providing support during the development of HEV (Wipke et al. 1999). This software was designed to execute performance analysis of different power trains. It simulates the flow of energy among the most important components of a vehicle as a set of discrete steps (Wipke et al. 1999). When a simulated vehicle follows a pre-defined driving cycle, the main outputs of the model are total energy consumption, amount of tailpipe emissions produced (specifically particulate Matter) and the estimated SOC of the battery. ADVISOR can also provide technical support regarding the vehicle's behavior under given velocity and acceleration conditions (Wipke et al. 1999). ADVISOR was developed completely in the MATLAB/SIMULINK® environment, which enables the integration of diverse technical models as blocks. A Graphical User Interface (GUI), seen in Fig. 3.4, is provided to the user, enabling the configuration of different power-trains and the selection of key parameters within the components. While the software itself does not provide routines to perform data and visual analytics, the MATLAB® environment is capable of providing support in this regard.

The Greenhouse Gases, Regulated Emissions, and Energy use in Transportation (GREET™) model was developed by the Argonne national laboratory. The main objective of the software is to evaluate and compare energy consumption and carbon-related emissions of transportation fuels and power trains from a life cycle perspective (Burnham et al. 2006). GREET™ consists of two modules as shown in Fig. 3.5: (i) GREET1 estimates the WTW energy consumption and production of emissions for the production of transportation fuels. It executes an evaluation alternative fuels pathway by modelling energy and feedstock consumption on a process level and/or providing emissions factors for particular combustion and energy production processes and (ii) GREET2 calculates cradle to gate energy and emissions of the production of vehicles and components. This is roughly done through estimating the total vehicles body-in-white (BIW) weight and breaking it down to its components weight. Afterward, this list is linked to a database that contains each material's energy

[5] NREL is a laboratory of U.S. Department of Energy.

	Wipke et al. 1999	Burnham et al. 2006	Menenga 2014	Hofer 2014	Bauer 2015	Onat et al. 2015	Egede et al. 2016	Bohnes et al. 2017	Wu 2018	Cox et al. 2018	Majeau-Bettet et al. 2011	Dunn et al. 2012	Ellingsen et al. 2014	Nelson et al. 2011	Schünemann 2015	Schönemann 2017	Spangenberger 2018	Sacchi et al. 2019	Haupt et al. 2018	Mery et al. 2013
Scope and system boundaries																				
Focus on environmental impacts of traction batteries	◔	◔	◔	◔	◔	◔	◔	◔	◔	◔	●	●	●	●	●	●	●	○	○	○
Focus on Life Cycle oriented environmental impact	○	◕	◑	◑	●	◑	◑	●	◑	◑	◑	◑	◑	○	◔	◔	◔	●	●	●
Focus on computational modelling	◕	●	●	◑	◔	○	◑	◑	◑	○	○	○	○	●	◑	●	●	◑	●	●
Raw materials	○	◑	○	◔	◔	◔	◔	◔	◔	◔	◑	◑	◔	●	◔	◔	◑	◔	○	○
Manufacturing	○	◑	◔	◑	◑	◔	◑	◑	◔	◔	◔	◕	●	●	●	●	◔	◔	○	○
Usage phase	●	●	●	●	●	●	●	●	●	●	○	○	◔	◔	○	○	○	●	○	○
End-of-Life	○	◑	○	○	◔	○	◔	◔	○	◔	◔	◑	◔	○	○	○	●	○	●	●
Assement criteria																				
Comprehensiveness of the assessment scope	◔	◕	◑	◑	◑	◑	○	◑	◑	◑	◑	◑	●	●	◑	◑	●	◑	●	◑
Model Usability / Life Cycle Inventory usability	◑	◑	◑	◑	◑	◑	○	◑	◑	◑	●	◑	●	●	●	◑	●	◕	◕	◑
Consideration of variability in the foreground system	●	◕	●	●	●	●	◑	◑	●	◑	●	●	●	●	◑	●	●	●	●	●
Consideration of variability in the background system	○	◑	◑	○	●	◑	◑	●	●	●	○	○	◑	○	○	◑	○	◑	◑	◑
Consideration of spatial variability	○	◔	○	○	◑	◑	●	◔	●	◑	○	○	○	○	○	○	○	●	◑	○
Modelling criteria																				
Integration of informatics tasks	○	○	○	○	○	○	○	○	○	◔	○	○	○	○	○	◑	○	◔	○	○
Visual and data analytics	◑	○	◑	◑	○	○	●	○	●	●	○	○	○	◑	◑	◑	◑	◑	◑	◑
Statistical methods	◑	◔	◑	○	◑	○	○	◕	●	●	○	○	○	◑	◑	◑	◑	◕	◑	◑
Prospective modelling	○	◑	○	●	●	○	◑	●	●	●	○	○	○	○	○	○	○	◕	●	◑
Scientific and empirical models	◔	◔	○	◑	○	○	◑	○	●	◑	○	○	○	◑	○	●	◑	◑	◑	●

Fig. 3.5 Qualitative evaluation of the selected approaches[6]

[6] Results are presented in terms of Harvey balls. A full ball (•) means that the criteria was fully fulfilled, partially filled balls (◔, ◑ and ◕) represent a partial fulfillment of the criteria. A blank ball (○) means that the criteria was not fulfilled.

intensity in units of energy per unit of mass. The latest version of the software enables modelling different vehicle technologies (Burnham et al. 2006). GREET™ is implemented in a total of 34 Microsoft® Excel™ spreadsheets with a GUI in Visual Basics for application. The logic of the software is fed with data mostly coming from repositories of emission inventories from the Environmental Protection Agency EPA, the international energy agency IEA, peer reviewed publications on current technology, and other results from third party process and driving cycles simulation tools. The outputs of the GREET™ environment consist of accounted total energy consumption classified into fossil, petroleum, natural gas and renewable energy. Additionally, the software lists the total amount of GHG emissions (CO_2, CH_4, N_2O, and CO_2-eq) and other transportation relevant emissions such as VOC, CO, NO_x, PM_{10}, $PM_{2,5}$ and SO_x, which, in the software, are divided into total and urban emissions.

The tool-based methodology presented by Mennenga (Mennenga 2014) was developed with the objective of providing support during the planning process of commercial fleets. The concept consists of seven modules implemented in Microsoft® Excel™ and VBA. For the modelling of the use phase of the fleets, an agent base simulation was implemented using AnyLogic™. This tool enables an initial set up of a fleet, consisting of a mix of conventional and alternative powertrains (Mennenga 2014). This configuration is further simulated within a specific scenario, in with important influencing factors such as electricity mix and a predefined usage profile. Further, the tool executes a life cycle-oriented evaluation of costs and environmental impacts using the results of the simulation, and the results are made available to the user in order to adjust the defined fleet. This life cycle-oriented approach additionally provides management support in the form of a dashboard with visuals depicting contribution analyses, trends and optimization potentials.

Within the context of THELMA project,[7] Hofer (2014) developed an integrated assessment framework for the comparison of modern vehicle technologies considering their economic, environmental and technical performance. The developed model integrates: (i) models covering both conventional and electric powertrains, (ii) different vehicle configurations in terms of their performance requirements such as range and size, (iii) different energy sources and future scenarios which, among other things, enables the consideration of changes in performance costs, performance, and energy prices and (iv) interdependencies among different vehicles configurations. This methodology does not provide detailed information regarding the data collected and made available vis a vis the modelling of the battery system material composition and manufacturing. It does, however provide some analytical models for analysis modelling and fleet simulation, as well as sensitivity analysis and optimization. The system is implemented in MATLAB® using the software ADVISOR™, and a web-based GUI is provided in order to interact with the different scenarios.

---

[7] Technology-centered Electric Mobility Assessment (THELMA) was a project from the Swiss Federal Institutes of Technology funded by Swiss Electric Research, the Competence Center for Energy and Mobility and the Swiss Erdöl Vereinigung. More information can be found at: http://www.thelma-emobility.net/.

Another work within the THELMA project was presented by Bauer and colleagues (2015). They developed an integrated scenario analysis framework for the analysis of current and future passenger vehicle technologies. In their approach, the authors classified the modules into endogenous and exogenous technology influences. This modelling approach integrates modules such as power-trains (conventional (ICEV) and alternative (HEV, BEV, PHEV, BEV, FCV and FCHEV), fuel types (gasoline, diesel, CNG, hydrogen), operating conditions (urban, rural, driving region, weather, etc.), future scenarios (electricity mixes, oil prices, etc.) and composition of the primary energy source. For the use phase, a simulation of different driving cycles was performed applying the software ADVISOR™ previously described. There is no further information regarding the computational implementation of the approach, and the tool was not provided as part of the project's publications. As with the previous contributions, the LCA module was integrated as a database of previously calculated inventories of aggregated LCIA results that is linked to the accounted material and energy flows.

The work done by Onat and colleagues (Onat 2015; Onat et al. 2015, 2016) introduced an integrated framework for the sustainability assessment of the U.S. transportation sector. This model enables the consideration of temporal and spatial variations in the assessment of the environmental impacts of electric-powered drive trains. Although it includes broad sub-models depicting the whole life cycle of the vehicles considered, the model's focus is predominantly on the analysis of the use phase. One particular characteristic of this model is the integration of system dynamic models to quantify the effect of complex interactions between social, economic and technological systems according to given policy scenarios. In their research, five different vehicle technologies are compared on the basis of their energy consumption and their production of GHG emissions in 50 U.S states. Additionally, three different scenarios are modelled: (i) average electricity mix, (ii) marginal electricity mix and (iii) 100% solar energy. For the estimation of the vehicle's life cycle (including components), as well as for the life cycle of the fuels modelled for the ICEV, the GREET™ model was used. There is no information available regarding the implementation of the model in a usable tool.

Egede (2016) presented an integrated modelling approach for the assessment of electric vehicles particularly focused on lightweight design aspects. Her methodology aims to provide a consistent framework for the life cycle comparison of the environmental impact of lightweight electric vehicles and conventional vehicles. A particularly important aspect of the concept introduced by Egede is that it enables the consideration of a large set of influencing factors within the spatial context of the usage stage. In order to do so, a database containing geographical information is linked to the models in the foreground system, depicting the energy consumed per kilometer for each of the technologies. The database integrates individual driving profiles, seasonal driving patterns, monthly minimum and maximum temperatures, sun rise and solar noon, as well as regional electricity mixes. Another important component of the concept introduced by Egede is the development of a visual communication instrument, called LCA world-maps, which acts as a medium to discern between the different technologies, regions and potential environmental impacts.

This enables the identification of the most appropriate power-train technology for a specific region and application. The methodology is implemented in Microsoft® Excel™, and the visualization is done using the computational statistics environment R. The estimation of the LCIA results is performed by linking pre-calculated datasets from commercial databases per unit of mass or energy. A user interface to the tool is not provided, and the approach does not integrate statistical analysis methodologies for the analysis of uncertainties and sensitivities.

Bohnes et al. (2017) developed an integrated assessment framework to evaluate the urban fleets deployment. Their approach proposes the use of life cycle assessment to quantify the environmental impact of transportation systems at a full fleet scale, including the consideration of future scenarios. The authors applied their framework to analyze the entire fleet of the city of Copenhagen during the period of 2016 to 2030, with a focus on the deployment of electric power train technologies. The four steps of the methodology were (i) definition of the focus of the study (city, timeframe, types of vehicles), (ii) definition of the scope of a vehicle-level LCA (e.g. types of power trains), (iii) definition of the scope of a fleet-level LCA (e.g. scenarios) and finally (iv) System modelling and data collection. The modelling approach integrates different models depicting fuel and electricity production, the vehicle's life cycle, charging infrastructure and some analytical models. An important contribution of this framework is the consideration of time differentiated foreground and background systems, along with temporal variations in the systems providing charging infrastructure and the development of electricity mixes. Additionally, the framework looks at variation regarding important factors such as the weight of the vehicle and the battery system. While the model is comprehensive in terms of scope, the models are quite generic and do not provide a high technological resolution. For instance, the battery system is examined in terms of its energy density and weight, but this examination does not detail its real material composition and its possible temporal development of the technology in terms of battery chemistries, emerging materials, supply chains among other factors. Nevertheless, routines to evaluate the sensitivity of important parameters are integrated in the model. Besides the fact that the LCA modelling was performed using the commercial software SimaPro®, there is no further information on the implementation of the tool, its usability and integration capabilities.

A similar approach is presented by Wu (2018). Wu developed an integrated assessment approach for the assessment of environmental impacts and TCO of lightweight electric vehicles. The purpose of this methodology is to estimate the cost-optimal fleet configuration able to achieve established environmental targets and, furthermore, to evaluate the influence of modern and future urban transportation schemes, such as such as ride-hailing and car-sharing. In a similar manner as Egede, Wu integrated regional differences across all U.S. cities, including climate, electricity-mixes and driving patterns. Additionally, the work presented by Wu includes a fleet stock evolution model that enables a dynamic assessment of the changes in the composition of a given fleet. Due to these unique features, this methodology allows for a spatiotemporal assessment of the environmental influences caused by the penetration of new power train and car-body technologies and/or mobility concepts over time.

No information regarding the implementation of the methodology and the execution of the LCA is provided.

The work done by Cox and colleagues (2018) present a comprehensive study in which a comparison of the environmental impacts of current and future transportation systems is performed. Although the publication does not define any assessment framework, the approach follows an integrated modelling sensitive to variations in both the background and the foreground systems. An important aspect of the modelling approach presented by the authors is the integration of external energy system models of future energy scenarios, while at the same time programmatically adjusting the background system datasets from ecoinvent to build datasets with future life cycle inventories. This enables the authors to consider a temporal differentiated electricity mix, while also considering the development of increased energy efficiency in alternative power-train technologies during the use phase of the vehicles. The authors also integrate statistical models for the assessment of global sensitivity analysis and variability, and presented an assessment of the most important factors affecting the environmental impact of EVs in the future. The implementation of the model was performed in Python and the LCA calculations were executed using the LCA python package Brightway (Mutel 2017). While the models were provided in the publication, they have a low usability as they were not included as a usable tool, but rather for the purpose of peer-reviewing the results presented in the paper.

Specific tools and methodologies for the assessment of the environmental impacts of traction batteries are rather scarce within the reviewed literature. The work presented by Majeau-Bettez and colleagues (2011) compared the environmental impacts of Nickel Metal Hybrid batteries and Li-ion batteries for electric vehicles. Although this research does not provide a tool or a specific modelling framework, their contribution provided one of the most comprehensive and transparent life cycle inventories currently available within the literature. This inventory has been used in a large number of other LCA studies and has been integrated as lineal sub-models into several tools. In their approach, Majeau-Bettez and colleagues linked a set of logical models to develop the inventory. First, the mass share of active and non-active components was fixed based on literature data. Next, the mass ratios of positive and negative electrodes were defined in order to obtain what the authors called "realistic energy performances," and the mass proportions between both electrodes were selected in order to have an identical reversible charge capacity. The mass of the rest of the cell was simply divided into the rest of the components. The energy use during the manufacturing phase was not modelled but estimated, following a top-down approach from industry and academic data.

The ANL has made great efforts to increase the resolution of the data used in the GREET™ model. As an example, the report provided by Dunn and colleagues (2014) in which the key cradle to gate (C2G) production processes of important cathode materials such as NMC, LFP and LCO, were reviewed and inventoried in order to build the materials and energy flows of the supply chain of battery systems. This information is currently available as part of the models in the GREET2 as reviewed before.

Ellingsen and colleagues (2014) contributed as well with a very comprehensive study for which a life cycle inventory was fully reported. The authors used data from the bill of materials from a real manufacturer to account for the exact amount and masses of components contained in a battery system. For the definition of the mass distribution at the cell level, the authors used the inventory from Majeau-Bettez and colleagues (2011) and the estimation of the energy consumption was done following a top-down approach from a pilot manufacturing plant. While the data presented in their paper is specific for a type and size of battery, their results have been included in plenty of other LCA studies. Several tools and assessment frameworks for the evaluation of battery pack costs have been developed in the last decade, though the lack of tools to evaluate their environmental implications remains. While the main focus of the present work is not the assessment of costs, the following approaches present important information regarding energy and material consumption. For this reason, these approaches have often been used to perform LCAs of battery systems.

One the most widely used tools is BatPac™ (Nelson et al. 2011). BatPac™ is a bottom-up model that enables the virtual design of a battery systems using different cathode and anode materials. Its objective is exclusively the estimation and analysis of manufacturing costs. One of the most important properties of BatPac™ is that the model enables for the consideration of physical limitations within the design of battery systems (e.g. geometry limitations and performance limitations). The tool is developed in such a way that the user has significant flexibility regarding parameter selection and variation. Inputs to the tool are classified into: (i) cell chemistry measured properties such as electrode porosity and specific energy capacity per unit of weight, (ii) pack requirements such as vehicle range, number of cells, required power and (iii) physical constraints such as max electrode thickness, cell potential, peak power and cell geometry. The outputs of this part of the model are, among other, volume and mass of the battery pack and its components, specific energy and power, and a list of materials required. BatPac™ additionally includes a hybrid bottom-up and top-down approach to estimate manufacturing costs. From the bottom-up, the model defines the requirements in space and the machines of a virtual factory, and following a top-down approach it estimates the amount of manufacturing energy required per unit of energy capacity of the battery cells. Hindering its usability and readability, BatPac™ has been completely implemented in Microsoft® Excel™, where logic, data and results are mixed, making it difficult to expand.

Schünemann (2015) also developed bottom-up model for the assessment of manufacturing costs of traction battery cells. The tool, called BaZeKaMo, integrates logic modules describing factory and manufacturing processes as well as cell properties. The model chains the modules inversely, allowing a connection between cell architecture and the specific cell production processes. It receives input parameters from three different categories: (i) cell design parameters such as cell chemistry, geometry and some performance KPIs, (ii) production parameters such as processing times, rate of cuttings, technical availability of the machines and (iii) cost-related parameters, such as personal costs, energy costs, and material prices. The outputs of the tool, including charts and analysis shown, are designed to support aspects of factory planning such as required production area, personnel requirements, energy consumption requirements and, for cost planning, productions costs, cost structures, volume of investments, total cost of ownership, and so on. In this way, BaZeKaMo is able to generate material, energy and cost flows for a whole chain of manufacturing processes and to link this information to a specific cell design. The data used to validate most of the processes comes from real manufacturing facilities, which makes the model highly realistic. The approach developed by Schünemann makes the dependency between cell properties and manufacturing process more transparent. The tool is also implemented using Microsoft® Excel™.

A further approach for the assessment and analysis of costs and environmental impacts during the manufacturing stage of battery cells is presented by Schönemann (2017). Schönemann developed a multi-scale simulation environment that enables the consideration of a large quantity of process, factory and product parameters and their interdependencies within a consolidated computational environment. This allows to create an enhanced interdisciplinary system understanding that provides a more robust support to product and production planners. While the model does not execute a LCA, it can be used to generate series of LCIA results that can be linked to LCA calculation routines. The approach is implemented paired with models done in AnyLogic ©, MATLAB® and Dymola.

Finally, a further approach, ReCell™ (Spangenberger 2018) has been developed by ANL with the objective of providing a platform to evaluate different recycling pathways for traction batteries and the environmental impacts rising from these pathways. The tool consists of three main modules: (i) a manufacture module containing only virgin materials and with a link to the BatPac™ tool previously described, (ii) a battery recycling model and (iii) a battery manufacture model containing only recycled materials. The recycling module is composed of sub-models describing transportation and collection processes, disassembly, cell recycling processes such as pyro-metallurgical/hydrometallurgical, as well as new cathode production processes. Inputs of the models are a definition and characterization of the stream of spent batteries, as well as key parameters of the selected recycling processes. The outputs of the model are, to name a few, the revenues, emissions, and waste, as well as flows of energy, materials and costs. The implementation of the tool was done completely in Microsoft® Excel™. The LCA results are calculated through a link to the GREET™ software.

3.2.3 Contributions Outside the Field of Electromobility

Two contributions outside of the scope of this work were identified as potentially enhancing the overview of the current state of research in terms of computational modelling in LCE. Mery and colleagues (2013) developed EVALEU, an integrated and flexible process modelling LCA tool that aims to integrate design models and LCA routines for the analysis of water treatment technologies. The work was motivated by the fact that water treatment plants are composed of a large set of unit processes, presenting a high variability in operation conditions which, in turn, depend on a large set of water quality and properties inputs. This situation causes the solutions range to be largely unconstrained, leading to high variability in the results of the LCA. EVALEAU was developed in Umberto® and integrates python scripts within the unit processes. It consists of a library of Unit Process Modules, which are basically parameterized Python™ scripts. These are, in turn, linked to model scripts depicting water chemistry properties. A water quality database contains raw water average quality data, case-specific/user-defined water quality data and drinking water standards. This data base is done in Excel™ and links to the foreground system modelled in Umberto®. A user interface is not available; however, the user is able to input predefined values using live links via Excel™. The background system is linked to ecoinvent. Am especially relevant aspect of this tool is the integration of a sensitivity analysis toolbox, which enables the development of Morris graphs and delivers engineering design data.

Sacchi and colleagues (2019) present an integrated framework for the assessment of wind turbines. The modelling approach allows for the consideration of diverse technologies, as well as for geographies and temporal aspects. The modelling approach consists of four modules. First, a database of fleet registries of wind turbines, with relevant information such as brand, model and location coordinates. This is matched to data regarding physical properties of each of the turbines. The second module contains models that characterize the spatial context of where the turbines are located. For instance, the model expresses the location of a given turbine in terms of its distance to the national electricity grid in order to estimate infrastructure requirements such as cables, while also taking into consideration weather properties. The third module refers to the foreground system. Here, a life cycle inventory in terms of mass, size, manufacturing requirements, installation requirements and disposal requirements is built and linked to the background database used in the model. Finally, a fourth module contains information regarding the amount of electricity produced by all of the wind turbines modelled. This lets the model estimate the lifetime of the turbines and the total service provided. The model provided integrates statistical tools for the execution of sensitivity analysis, and is able to produce some visual analytics aids. The model is implemented in Python and connected to Brightway (Mutel 2017) to solve the LCAs for each of the turbines modelled in the specific scenarios.

3.2.4 *Evaluation of Approaches and Summary of Findings*

It cannot be argued that there is a consensus regarding the environmental impact of traction batteries. The chain of stakeholders taking part in one form or another of the emerging electromobility field are still largely uninformed about the potential consequences of this technological change. Although this transition might bring essential environmental repercussions, the current state of research on the environmental assessment of traction batteries still lacks clarity. There are numerous scientific publications in this field, but the assumptions made and the data used has been found to be largely based on no more than five contributions. As a result, the available results of LCA for electromobility are characterized by being very uncertain and sensitive to a large set of parameters. As found in the literature, essential missing information include the manufacturing energy; material composition of the battery systems; material production processes; battery aging and lifetime; battery recycling processes and battery recyclability; and type and composition of the battery management system, to name a few (Ellingsen et al. 2017; Peters et al. 2017). While there have been some efforts to consolidate the existing published inventories to make them useful and comparable (Peters and Weil 2018), the information gap remains, as these inventories are usually not interoperable. This hampers their comparability and to provide engineering support engineering support. In addition, the available databases are not comprehensive enough to provide a geographical, technological and temporal resolution to provide the higher resolution data necessary. This is particularly when considering all of the processes involved in the extraction and mining of minerals and production of really relevant materials (copper, cobalt, nickel, manganese, etc.)

As seen in Chap. 2, a large and diverse group of professional backgrounds take part in the development of traction batteries. This ranges from material discovery and development, through to chemistry, electrical engineers, manufacturing engineers, and many others. Notwithstanding, the LCA literature on traction batteries (meaning the available knowledge on the environmental effects of traction batteries and therefore of EVs) has been largely developed and made available by professionals in fields such as industrial ecology and LCA, and not by battery experts. A rigorous and detailed modelling of the life cycle of a battery system is a highly complex task for which LCA practitioners are not always qualified, and for which the integration of professional knowledge from different disciplines is required.

In spite of that, there are not specific tools and modelling approaches for the assessment of the environmental impacts of traction batteries. Table 3.1 summarizes the approaches selected and reviewed, and presents a qualitative evaluation based on the criteria previously defined. The first most evident conclusion of the analysis presented in Table 3.1 is the fact that there are tools or tool-based methodologies for the environmental assessment of traction batteries. While some tools like BatPac™ (Nelson et al. 2011) and other cost models mentioned in the above review are capable of estimating material composition and the performance of battery systems, these are focused on the assessment of costs.

As observed in Fig. 3.5, there is a disconnect between EV-based studies and traction batteries-based studies. Almost all of the modelling approaches and tools aiming at estimating the life cycle impacts of EVs are mainly focused on modelling the usage stage. Tools such as Advisor™ (Wipke et al. 1999) have been traditionally applied to the estimation of emissions from a TTW perspective. Later on, this tool was integrated into modelling approaches considering full fuel and vehicles cycles. In the literature reviewed, the usage stage is usually considered in terms of energy consumed per kilometer. While this stage has been traditionally analyzed on a tank to wheel perspective, there is an increasing trend within this scientific community to understand the influence of the spatial contexts in terms of electricity mix, weather and driving patterns at a vehicle level (Egede 2016) and fleet level (Mennenga 2014;

Onat 2015; Wu 2018; Bohnes et al. 2017). Furthermore, comprehensive tools such as GREET™ have expanded their capabilities and scope by integrating not only the possibility to analyze the vehicle's life cycle, but also by extending the system boundaries to the modelling of the use stage of a vehicle in specific regions in the U.S.

In these models, the battery system is often considered as a black box and modelled using LCIs from openly published studies such as (Majeau-Bettez et al. 2011; Ellingsen et al. 2014; Notter et al. 2010; Dunn et al. 2015b). These approaches are mainly interested in the evaluation of greenhouse gases, and seldom evaluate other environmental impacts.

In turn, the modelling approaches which model and assess traction battery systems are mostly cradle to gate studies, with the exception of some studies focusing on the recycling phase. These are mostly focused on geometrical models or mass distribution models, but often do not integrate electrochemistry-based mass-energy models into their approach. Modelling approaches that look at costs are focused on both material distribution models and manufacturing requirements in terms of energy requirements. The manufacturing stage has been the focus of several modelling approaches, with the goal of the modelling costs. These studies are nevertheless based on assumption that, due to confidentiality issues, are difficult to follow.

Models such as the ones presented by Schönemann (2015, 2017) and the BatPac™ model provide flexible and extensible process and factory models that can be further implemented in integrated LCE approaches.

While the importance of considering the specific production routes of key materials such as cobalt and nickel has been often acknowledged (Schmidt et al. 2016; Ahmed et al. 2017), the modelling of the production of raw materials in LCA of traction batteries remains, so far, largely neglected. The lack of models and data becomes increasingly evident the deeper the analysis goes into the material supply chain for traction batteries, and the thus its associated environmental impact. Other modelling approaches from the field of industrial ecology, such as multiregional environmental input output analysis MRIO EIO, have also been applied as an effort to quantify these implications (Afshar 2017; Sen et al. 2018). However, this approach does not provide enough insights at a product level to support engineers.

Regarding the reviewed modelling paradigms of the existing approaches in the field of electromobility, important conclusions can be drawn. As in almost all of the contributions, the development of a computational tool was not the main focus. None of the approaches reviewed are capable of integrating informatics tasks such as data management and linking. This has some consequences regarding the implementation of visual analytics methodologies and statistical packages for the analysis of sensitivities and uncertainties.

Some exceptions in this regard are the contributions from Cox et al. (2018), Wu (2018), whose approaches, due to the fact that their tools are developed in scripting languages, offer the possibility to integrate sensitivity analysis packages as well as well-known visualization packages. Additionally, the same situation is presented regarding the integration of empirical or scientific models. The possibility of performing prospective modelling is enabled by few of the approaches reviewed.

This is particularly an issue given in those contributions focused on the analysis of EVs and having as a scope the estimation of future environmental consequences. Prospective modelling is regularly done in terms of a scenario modelling, integrating technological changes in the foreground system and factors of the spatial context, such as the electricity mix. To sum up, current modelling approaches in LCE do not enable a robust and fast coverage of the important parameters influencing the environmental impact of traction batteries and electric vehicles. The manufacturing phase will keep evolving and being optimized, and new materials will emerge, making existing LCA models and inventories obsolete. The rigid modelling approach found within the LCA community in general and in the field of electromobility in particular, hampers the swift discovery of knowledge which could result in a better-informed technology roadmapping, a better product and a better process design.

The field lacks a modular modelling approach, which enables a fast examination of plenty of product systems in the field of electromobility and traction batteries, while being extendible and multidiscipline-oriented as well able to carry out analytical functions. Such an approach would cope with the current trend towards a higher computational flexibility in other fields and generate a new awakening in the way life cycle assessment software and databases are being planned and developed.

References

Afshar S (2017) Eco-efficiency of electric vehicles in the United States: a life-cycle assessment based principal component analysis. earsiv.sehir.edu.tr

Aguirre K, Eisenhardt L, Lim C et al (2012) Lifecycle analysis comparison of a battery electric vehicle and a conventional gasoline vehicle 1–33

Ahmed S, Nelson PA, Gallagher KG et al (2017) Cost and energy demand of producing nickel manganese cobalt cathode material for lithium ion batteries. J Power Source 342:733–740. https://doi.org/10.1016/j.jpowsour.2016.12.069

Amarakoon S, Smith J, Segal B (2013) Application of life- cycle assessment to nanoscale technology : lithium-ion batteries for electric vehicles 1–119

Bartolozzi I, Rizzi F, Frey M (2013) Comparison between hydrogen and electric vehicles by life cycle assessment: a case study in Tuscany, Italy. Appl Energy 101:103–111. https://doi.org/10.1016/j.apenergy.2012.03.021

Bauer C, Hofer J, Althaus HJ et al (2015) The environmental performance of current and future passenger vehicles: life cycle assessment based on a novel scenario analysis framework. Appl Energy 157:871–883. https://doi.org/10.1016/j.apenergy.2015.01.019

Bohnes FA, Gregg JS, Laurent A (2017) Environmental impacts of future urban deployment of electric vehicles: assessment framework and case study of Copenhagen for 2016–2030. Environ. Sci. Technol. https://doi.org/10.1021/acs.est.7b01780

Burnham A, Wang M, Wu Y (2006) Development and applications of GREET 2.7—the transportation vehicle-cycle model. osti.gov

Cerdas F, Andrew S, Thiede S, Herrmann C (2018a) Environmental aspects of the recycling of lithium-ion traction batteries. In: Lithorec, pp 267–288

Cerdas F, Titscher P, Bognar N et al (2018b) Exploring the effect of increased energy density on the environmental impacts of traction batteries: a comparison of energy optimized lithium-ion and lithium-sulfur batteries for mobility applications. Energies 11:150. https://doi.org/10.3390/en11010150

Cox B, Mutel CL, Bauer C et al (2018) Uncertain environmental footprint of current and future battery electric vehicles. Environ Sci Technol 52:4989–4995. https://doi.org/10.1021/acs.est.8b00261
Cox BL, Mutel CL (2018) The environmental and cost performance of current and future motorcycles. Appl Energy 212:1013–1024. https://doi.org/10.1016/j.apenergy.2017.12.100
Cusenza M, Bobba S, Ardente F et al (2019) Energy and environmental assessment of a traction lithium-ion battery pack for plug-in hybrid electric vehicles. J Clean Prod 215:634–649. https://doi.org/10.1016/j.jclepro.2019.01.056
Deng Y, Li J, Li T et al (2017) Life cycle assessment of lithium sulfur battery for electric vehicles. J Power Sources 343:284–295. https://doi.org/10.1016/j.jpowsour.2017.01.036
Duce AD, Egede P, Öhlschläger G et al (2013) Guidelines for the LCA of electric vehicles. European Union Seventh …, Brussels
Dunn JB, Gaines L, Sullivan J, Wang MQ (2012) Impact of recycling on cradle-to-gate energy consumption and greenhouse gas emissions of automotive lithium-ion batteries. Environ Sci Technol 46:12704–12710. https://doi.org/10.1021/es302420z
Dunn J, James C, Gaines L, Gallagher K (2014) Material and energy flows in the production of cathode and anode materials for lithium ion batteries. Argonne Natl Lab
Dunn JB, Gaines L, Kelly JC et al (2015a) The significance of Li-ion batteries in electric vehicle life-cycle energy and emissions and recycling's role in its reduction. Energy Environ Sci 8:158–168. https://doi.org/10.1039/C4EE03029J
Dunn JB, James C, Gaines L et al (2015b) Material and energy flows in the production of cathode and anode materials for lithium ion batteries. osti.gov
Egede P (2016) Environmental assessment of lightweight electric vehicles 94–98
Egede P, Dettmer T, Herrmann C, Kara S (2015) Life cycle assessment of electric vehicles—a framework to consider influencing factors. Procedia CIRP 29:233–238. https://doi.org/10.1016/j.procir.2015.02.185
Ellingsen LAW, Majeau-Bettez G ... (2014c) Life cycle assessment of a lithium-ion battery vehicle pack. J Ind Ecol
Ellingsen LA-WW, Majeau-Bettez G, Singh B et al (2014) Life cycle assessment of a lithium-ion battery vehicle pack. J Ind Ecol 18:113–124. https://doi.org/10.1111/jiec.12072
Ellingsen LA, Singh B, Strømman AH (2016) The size and range effect : lifecycle greenhouse gas emissions of electric vehicles. Environ Res Lett 11:054010. https://doi.org/10.1088/1748-9326/11/5/054010
Ellingsen LAW, Hung CR, Strømman AH (2017) Identifying key assumptions and differences in life cycle assessment studies of lithium-ion traction batteries with focus on greenhouse gas emissions. Transp Res Part D Transp Environ 55:82–90. https://doi.org/10.1016/j.trd.2017.06.028
Faria R, Marques P, Moura P et al (2013) Impact of the electricity mix and use profile in the life-cycle assessment of electric vehicles. Renew Sustain Energy Rev 24:271–287
Faria R, Marques P, Garcia R et al (2014) Primary and secondary use of electric mobility batteries from a life cycle perspective. J Power Sources 262:169–177. https://doi.org/10.1016/j.jpowsour.2014.03.092
Gao L, Winfield ZC (2012) Life cycle assessment of environmental and economic impacts of advanced vehicles. Energies 5:605–620. https://doi.org/10.3390/en5030605
Girardi P, Gargiulo A, Brambilla PC (2015) A comparative LCA of an electric vehicle and an internal combustion engine vehicle using the appropriate power mix: the Italian case study. Int J Life Cycle Assess 20:1127–1142. https://doi.org/10.1007/s11367-015-0903-x
Hawkins TR, Gausen OM, Strømman AH (2012) Environmental impacts of hybrid and electric vehicles-a review. Int J Life Cycle Assess 17:997–1014. https://doi.org/10.1007/s11367-012-0440-9
Hawkins TR, Singh B, Majeau-Bettez G, Strømman AH (2013) Comparative environmental life cycle assessment of conventional and electric vehicles. J Ind Ecol 17:53–64. https://doi.org/10.1111/j.1530-9290.2012.00532.x

Hofer J (2014) Sustainability assessment of passenger vehicles: analysis of past trends and future impacts of electric powertrains. research-collection.ethz.ch

Huo H, Cai H, Zhang Q et al (2015) Life-cycle assessment of greenhouse gas and air emissions of electric vehicles: a comparison between China and the U.S. Atmos Environ 108:107–116. https://doi.org/10.1016/j.atmosenv.2015.02.073

Kim HC, Wallington TJ, Arsenault R et al (2016) Cradle-to-gate emissions from a commercial electric vehicle li-ion battery: a comparative analysis. Environ Sci Technol 50:7715–7722. https://doi.org/10.1021/acs.est.6b00830

Kwade A, Diekmann J, Hanisch C et al (2016) Recycling von lithium-ionen-batterien—lithoRec II Abschlussberichte der beteiligten Verbundpartner. Verbundprojekt im Rahmen des Förderprogramms. "Erneuerbar Mobil" des Bundesministeriums für Umwelt, Naturschutz, Bau und Reakt

Larcher D, Tarascon J (2014) Towards greener and more sustainable batteries for electrical energy storage. Nat Publ Gr 7. https://doi.org/10.1038/nchem.2085

Li B, Gao X, Li J, Yuan C (2014) Life cycle environmental impact of high-capacity lithium ion battery with silicon nanowires anode for electric vehicles. Environ Sci Technol 48:3047–3055. https://doi.org/10.1021/es4037786

Li W, Stanula P, Egede P et al (2016) Determining the main factors influencing the energy consumption of electric vehicles in the usage phase. In: Procedia CIRP. Elsevier, pp 352–357

Majeau-Bettez G, Hawkins TR, Strømman AH (2011) Life cycle environmental assessment of Li-ion and nickel metal hydride batteries for plug-in hybrid and battery electric vehicles. Environ Sci Technol 45:4548–4554. https://doi.org/10.1021/es103607c (Supporting Information)

Mennenga MS (2014) Lebenszyklusorientierte Flottenplanung mit alternativ angetriebenen Fahrzeugkonzepten

Mery Y, Tiruta-Barna L, Benetto E, Baudin I (2013) An integrated "process modelling-life cycle assessment" tool for the assessment and design of water treatment processes. Int J Life Cycle Assess

Mutel C (2017) Brightway : an open source framework for Life Cycle Assessment. 47:11–12. https://doi.org/10.21105/joss.00236

Nealer R, Hendrickson TP (2015) Review of recent lifecycle assessments of energy and greenhouse gas emissions for electric vehicles. Curr Sustain Energy Reports 2:66–73. https://doi.org/10.1007/s40518-015-0033-x

Nelson P, Gallagher KG, Bloom I, Dees DW (2011) Modeling the performance and cost of lithium-ion batteries for electric-drive vehicles chemical sciences and engineering division. Argonne Natl Lab 116. https://doi.org/10.3133/fs20143035

Nordelöf A, Messagie M, Tillman AM et al (2014) Environmental impacts of hybrid, plug-in hybrid, and battery electric vehicles—what can we learn from life cycle assessment? Int J Life Cycle Assess 19:1866–1890

Nordelöf A, Poulikidou S, Chordia M et al (2019) Methodological approaches to end-of-life modelling in life cycle assessments of lithium-ion batteries. Batteries 5. https://doi.org/10.3390/batteries5030051

Notter DA, Gauch M, Widmer R et al (2010) Contribution of li-ion batteries to the environmental impact of electric vehicles. Environ Sci Technol 44:6550–6556. https://doi.org/10.1021/es903729a

Onat NC (2015) Integrated sustainability assessment framework for the US transportation. etd.fcla.edu

Onat NC, Kucukvar M, Tatari O (2015) Conventional, hybrid, plug-in hybrid or electric vehicles? State-based comparative carbon and energy footprint analysis in the United States. Appl Energy 150:36–49. https://doi.org/10.1016/j.apenergy.2015.04.001

Onat NC, Kucukvar M, Tatari O, Egilmez G (2016) Integration of system dynamics approach toward deepening and broadening the life cycle sustainability assessment framework: a case for electric vehicles. Int J Life Cycle Assess 21:1009–1034. https://doi.org/10.1007/s11367-016-1070-4

Otte E, Rousseau R (2002) Social network analysis: a powerful strategy, also for the information sciences. J Inf Sci

Peters JF, Baumann M, Zimmermann B et al (2017) The environmental impact of Li-Ion batteries and the role of key parameters—A review. Renew Sustain Energy Rev 67:491–506

Peters JF, Weil M (2018) Providing a common base for life cycle assessments of Li-Ion batteries. J Clean Prod 171:704–713. https://doi.org/10.1016/j.jclepro.2017.10.016

Rangaraju S, De Vroey L, Messagie M et al (2015) Impacts of electricity mix, charging profile, and driving behavior on the emissions performance of battery electric vehicles: a Belgian case study. Appl Energy 148:496–505. https://doi.org/10.1016/j.apenergy.2015.01.121

Sacchi R, Besseau R, Perez-Lopez P, Blanc I (2019) Exploring technologically, temporally and geographically-sensitive life cycle inventories for wind turbines: a parameterized model for Denmark. Renew Energy

Schmidt T, Buchert M, Schebek L (2016) Investigation of the primary production routes of nickel and cobalt products used for Li-ion batteries. Resour Conserv Recycl 112:107–122. https://doi.org/10.1016/j.resconrec.2016.04.017

Schönemann M (2017) Multiscale simulation approach for battery production systems

Schünemann J-H (2015) Modell zur Bewertun der Herstellkosten von Lithiumionenbatteriezellen

Sen B, Onat NC, Kucukvar M, Tatari O (2018) Material footprint of electric vehicles: a multiregional life cycle assessment. J Clean Prod. https://doi.org/10.1016/j.jclepro.2018.10.309

Spangenberger J (2018) ReCell: Closed-loop battery recycling model. Argonne Natl Lab 5543

Steubing B, Wernet G, Reinhard J et al (2016) The ecoinvent database version 3 (part II): analyzing LCA results and comparison to version 2. Int J Life Cycle Asses 3:1–13. https://doi.org/10.1007/s11367-016-1109-6

Tagliaferri C, Evangelisti S, Acconcia F et al (2016) Life cycle assessment of future electric and hybrid vehicles: a cradle-to-grave systems engineering approach. Chem Eng Res Des

Thomitzek M, Cerdas F, Thiede S, Herrmann C (2019) Cradle-to-gate analysis of the embodied energy in lithium ion batteries. Procedia CIRP 80:304–309. https://doi.org/10.1016/j.procir.2019.01.099

Thomitzek M, Von Drachenfels N, Cerdas F et al (2019) Simulation-based assessment of the energy demand in battery cell manufacturing. Procedia CIRP 80:126–131. https://doi.org/10.1016/j.procir.2019.01.097

Weidema BP, Bauer C, Hischier R et al (2013) Data quality guideline for the ecoinvent. Swiss Cent Life Cycle Invent 3

Wernet G, Bauer C, Steubing B et al (2016) The ecoinvent database version 3 (part I): overview and methodology. Int J Life Cycle Assess

Wipke K, Cuddy M, Bharathan D et al (1999) Advisor 2.0: a second generation advanced vehicle simulator for systems analysis. NAEVI. NREL/TP-540-25928

Wu D (2018) Spatiotemporal lifecycle assessment of the light-duty vehicle fleet of United States. Massachusetts Institute of Technology (2014)

Yuan C, Deng Y, Li T, Yang F (2017) Manufacturing energy analysis of lithium ion battery pack for electric vehicles. CIRP Ann Manuf Technol 66:53–56. https://doi.org/10.1016/j.cirp.2017.04.109

Zackrisson M, Avellán L, Orlenius J (2010) Life cycle assessment of lithium-ion batteries for plug-in hybrid electric vehicles-Critical issues. J Clean Prod 18:1517–1527. https://doi.org/10.1016/j.jclepro.2010.06.004

Chapter 4
Concept Development: Integrated Computational Life Cycle Engineering for Traction Batteries

Contents

Modelling, analyzing and understanding the life cycle environmental impacts of product systems is, in general, a time demanding and interdisciplinary task. It requires specialized LCA modelling skills and deep knowledge of the product system under analysis. EVs are particularly complex product systems. Their environmental impacts are sensitive to a large set of parameters in the foreground system (FS) as well as to their interaction within specific spatial and temporal contexts and background system (BS). Available commercial software tools fail to streamline the integration of this complexity into analytical models. This leads to the development of simplified models that hamper well-informed decision-making processes, therefore holding back the implementation of LCA within engineering activities. In addition, current approaches limit the application of visual analytics due to the lack of robustness and comprehensibility of the results achieved. To face these challenges, an Integrated Computational Life Cycle Engineering ICLCE[1] framework for EVs has been proposed (Cerdas et al. 2018). The ICLCE framework described in this chapter aims

[1] The description of the framework developed within this research has been published in the CIRP Annals of Manufacturing technology (Cerdas et al. 2018) and summarizes the content of this chapter. This implies that there are similitudes between the journal article and this section.

F. Cerdas, *Integrated Computational Life Cycle Engineering for Traction Batteries*, Sustainable Production, Life Cycle Engineering and Management,
https://doi.org/10.1007/978-3-030-82934-6_4

to support the fast and comprehensive modelling of complex FSs in the electromobility field, and their interactions with diverse backgrounds and spatial contexts. This chapter introduces the ICLCE framework developed for the analysis of the environmental impacts of EVs and their components, and its integration within product and process engineering activities. It includes a detailed description of the development of the concept, the involved elements, the practical implementations and the different user interaction cases.

4.1 Systems Perspective in ICLCE

One of the main motivations behind ICLCE for EVs is to provide a set of computational modules (see Fig. 4.1) that is able to quickly and efficiently consider the different systems exerting influence on the environmental footprint of an electric vehicle. Additionally, the ICLCE framework intends to support the interpretation of the results of a LCA.

As seen in Chap. 2, a product system depicting the life cycle of an EV is largely unconstrained. The product system (PS) in the ICLCE framework for EVs consists of every unit process in the technosphere, as well as the amount and quantity of elementary flows crossing the boundary to or from the biosphere. This includes unit processes linked through material and/or energy flows, and relevant spatial and temporal information influencing the nature and amount of elementary flows created. The unit processes of a PS are further classified into FS and BS processes. The

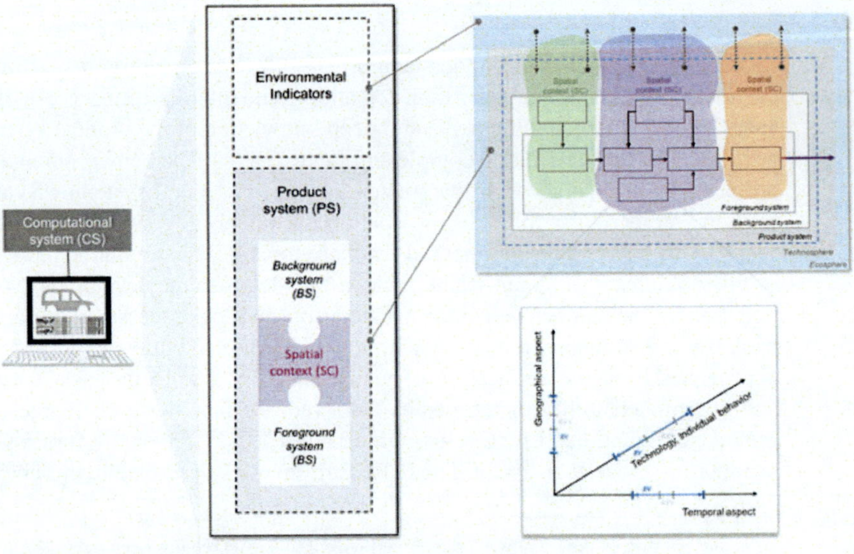

Fig. 4.1 The systems perspective in the ICLCE framework for electric vehicles

processes in the FS are specific to the electromobility's supply chain. Examples of foreground processes in this context are the manufacturing stage of battery cells, the driving operation during the usage stage, a recycling process to recover material from spent batteries and the production of active materials and their precursors. From an LCE perspective, the processes in the FS are largely modelled based on primary data. The related data are normally collected from research activities, and can be measured or generated from physical or mathematical models. In the context of this work, the output of a FS is called a product object (PO). The BS is comprised of processes in both the upstream and downstream supply chain of an EV that are not strictly specific to it. Examples of BS processes are the production of metal commodities such as copper and steel, the production of electricity and energy carriers, and transportation and waste management activities, to mention a few.

The processes in the foreground and background system are normally subject to a high variability and easily influenced by modelling choices. This variability is further multiplied by the often large influence of spatial and temporal factors. Some examples of such factors are: (i) geographical information such as weather conditions, (ii) information having an effect on the deployment of electric vehicles within specific markets, (iii) demographics (e.g. urban population density), (iv) composition of local electricity mixes, (v) inter-individual behavior affecting the energy consumption of the vehicle during the use stage (e.g. driving patterns and behavior, average commuting times and distances for the specific spatial context, etc.).

Although both the specific spatial contexts and the background system processes cannot be easily influenced by decision makers, the integration of this information into engineering activities might help to disclose a holistic understanding regarding the environmental impacts of the product system (Cerdas et al. 2018). This might further lead engineers to develop customized solutions, adjustments of designs or site-specific private mobility policies.

To consistently consider the inherent variability described, computational supported modelling approaches are essential. A computational system (CS) is, in this context, a playground of models and tools for engineers that help them extend their understanding regarding the consequences of their technological developments on systems thinking mindset. The CS synthesizes the information from both the technosphere and the biosphere as a set of environmental indicators that is linked to the provision of the electric vehicle's functionality.

4.2 Synthesis of Needs, Objectives and Requirements

As concluded in Chap. 3, there is currently a methodological gap in the field of LCE. A modelling approach that leads to a robust evaluation of the life cycle environmental impacts of product systems while considering a large variability in the configuration of the product system and enabling the use of this information to life cycle engineer traction batteries and electric vehicles is missing.

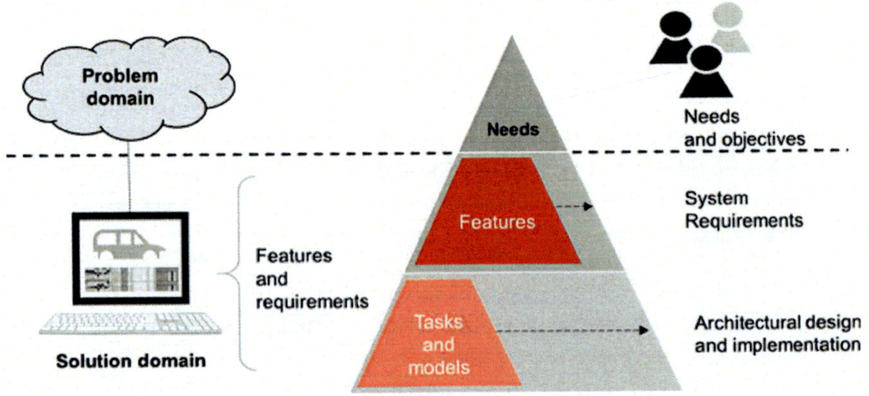

Fig. 4.2 Overview of the problem and solution domain to derive the concept. Adapted from Leffingwell and Widrig (2000)

The frameworks for the management and identification of software requirements (Fig. 4.2) developed by Leffingwell and Widrig (2000) has been adapted to map the needs and the requirements of the ICLCE concept. To derivate the required functionality, two domains, problem and solution, are described. In the problem domain, the needs of the different stakeholders/users (personas) of the system to be developed are gathered and described.

After having identified the needs, a list of high level required features is created which are afterwards broken down into a set of tasks and models. These are then divided into concrete tasks or specific models whose implementation is presented in Chap. 5. The collection of requirements, presented in the following sections, is not intended to be fully complete, as the requirements for an ICLCE might evolve demanding a different or enhanced functionality which is in turn linked to other themes and tasks. This is in essence an important characteristic of ICLCE, its ability to grow, update and evolve.

4.2.1 Synthesis of Needs

As seen in Chaps. 2 and 3, the scope of research in the field of electromobility is very broad. Relevant research questions can be found in each of the life cycle stages of an electric vehicle and its components. This encompasses product designs, materials, production processes, working principles of a specific technology, and business models affecting the way in which a technology is manufactured, used or disposed.

To summarize the needs upon which the ICLCE is to be developed, an analysis of the problem domain is firstly done. In this regard, three users were considered for the synthesis of the needs. First, ICLCE is to support *researchers and early developers* in the development of new materials, processes and products. These activities take place throughout the life cycle of a given product system. Next, ICLCE aims at providing *manufacturers* with a methodology and developed tool to support an integrated engineering of their products, taking in consideration specific environmental issues throughout the life cycle and the interactions between FS, BS and SC. As LCE becomes increasingly important within engineering activities, the ICLCE approach emerges as an essential tool to support the holistic modelling usually performed by the *LCE modeler*. The needs of these three groups of users is summarized in Table 4.1.

Table 4.1 Synthesis of needs of the users considered for the development of the concept

User	Needs
Researchers and (early) product developers	1. To quantify easily and fast the potential environmental impacts of the technology developed (material, process, component, etc.) on a life cycle perspective 2. To understand the linkage between life cycle environmental impacts and specific technical parameters in their fields 3. To be able to integrate LCA results into optimization routines in order to have more robust designs and products
Manufacturers	4. To quantify easily and fast the product's footprint for communication purposes 5. To have better decision support from the LCA tools (e.g. through visualization techniques and analysis of uncertainties) 6. To integrate easily upstream environmental information or models from suppliers to quantify specific footprints of products 7. To easily derive design and development constraints based on environmental aspects
Life cycle engineers	8. To have a modelling approach that consistently and efficiently builds product systems as a result of the interaction of coupled models 9. To generate large amount of results in order to ease the knowledge and insights discovery process 10. To interpret insights and give sound recommendations to the disciplines engaged in the respective life cycle stages

4.2.2 *Analysis of Requirements*

Based on the general objective *to develop a methodological framework to integrate computational modules that enable the life cycle engineering of traction batteries and electric vehicles*, the specific objectives have been structured following the four steps of the LCA framework. This is because the computational system to be developed is intended to support the modelling of product systems as guided by this framework and to support engineering activities and decision making. Table 4.2 lists the specific objectives of the ICLCE framework along with their respective requirements.

4.3 Framework and General Modelling Scheme

4.3.1 *Framework Development and Reference Architecture*

Integrated Computational Engineering and its opportunities for LCE

Engineering models are normally developed to represent phenomena in different disciplines (e.g. solid mechanics, fluid dynamics, chemistry) and length scales (i.e. from an atomic scale to a complex product such as a vehicle) (Horstemeyer 2012). The increasing computational power has led to a larger integration of modelling and simulation in engineering. A general, discipline-neutral, definition of Integrated Computational Engineering (ICEg) is given by Schmitz and Prahl (2012). By analyzing the unary and binary definitions of the terms "integrated", "computing" and "engineering" (see Table 4.3) they defined ICEg as any combination of computational engineering activities (i.e. models) in a functional unified system (Schmitz and Prahl 2012).

Particularly in the field of materials engineering, the inherent interaction between disciplines and the heterogeneity of scale-lengths often implied misunderstandings regarding modelling boundaries and interfaces (Horstemeyer 2012).

Materials Engineering (ME) is the field that aims at engineering the properties (e.g. grain size, phase fractions, crystallographic orientation, defects among others) of materials. These are mostly defined by the microstructures of the materials which is, in turn, the consequence of a particular chemical composition (Schmitz and Prahl 2012). Materials are seen as a system which is arranged in a multilevel structure with a given hierarchy of length scales. In this regard, the discipline of computational material design has been based on a multilevel structure of hierarchical and multidisciplinary design models (Olson 1997).

In spite of the differences of scope, both LCE and materials engineering (ME) are system analysis disciplines aiming to support product development. Both fields focus on the analysis of large, multi-scale and interdisciplinary systems with the goal of optimizing designs. To this end, both disciplines rely on the bottom-up modelling

Table 4.2 Synthesis of concept requirements

Goal and scope	
O1	*The concept shall enable a life cycle-oriented modelling disregarding what is to be defined as a foreground system*
R1	To provide the user an environment in which it is possible to integrate discipline specific models into a life cycle model consisting of models in other life cycle stages. The discipline specific model integrated will be defined by the user as the foreground system enabling therefore the detailed examination of that foreground system as a block
O2	*The concept shall not be limited to one specific type of environmental impact*
R2	to enable selecting from a list of impact assessment methodologies as it is usually done in commercial LCA software
O3	*The concept shall allow tracking, tracing and comparing the product systems built*
R3	A standard data structure and exchange format is required to be defined for the description of a product system. This format is required to be valid for each product system created and it is required to be uniquely identified
R4	To uniquely identify parts, components, flows and parameters within a product system is required
Life cycle inventory	
O4	*The concept should enable the user to integrate foreground system models*
R5	To enable the integration of models to describe physically and geometrically the architecture of an EV
R6	To enable the integration of top-down and/or process-based production models to build sub-inventories of emissions generated and material and energy demand during the raw-material production, manufacturing and recycling stages
R7	To enable the integration of energy demand models for electric vehicles in the use stage
R8	To enable the integration of battery aging models for the analysis of usage stage

(continued)

Table 4.2 (continued)

R9	To enable the consideration multidisciplinary parameters and phenomena from research fields distributed along the life cycle of a product system
O5	*The concept should enable exploring the effect of variating background systems*
R10	To enable applying changes in the background system by adjusting commercial datasets
R11	To enable modelling technological landscapes based on forecasts
O6	*To enable the consideration of the influence of the specific spatial contexts*
R12	To enable linking a product system to the respective regions which had an influence on it
R13	To enable the integration of spatial factors exerting an influence on the environmental impact of traction batteries and electric vehicles in each of the life cycle stages. This ranges from linking electricity mixes in time and place for each of the life cycle stages of an EV, through considering driver patterns and behavior in particular cities, to taking into consideration road networks and climatic conditions in the specific cities
O7	*To enable the flexible link of commercial databases*
R14	To allow linking to commercial disaggregated background LCI datasets without incurring in storing aggregated pre-calculated datasets into the model. ICLCE is required to link a product system to a commercial database and evaluate it
Life cycle impact assessment	
O8	*To evaluate a product system for a given list of environmental impacts for impact categories selected in a spatially differentiated manner*
R15	To perform environmental impact assessments for all impact categories otherwise available in commercial LCA software
R16	To enable linking emissions to the places where they originate
Interpretation	
O9	*To enhance decision making support*
R17	To deliver LCA make LCA results useful for engineers

(continued)

Table 4.2 (continued)

	The ICLCE framework is required to allow linking a product system to modified LCI datasets in a loop so as to enable sensitivity analysis in the background system
R18	To provide enhanced visualization methodologies in order to represent complexity in a more understandable manner while satisfying the goal and scope defined
R19	To integrate mathematical frameworks for the evaluation of both local and global sensitivity analysis
R20	To disclose product-production-environmental impacts dependencies in order to obtain better understanding about what depend on what
Implementation in a computational system	
O10	*To be implementable in an open source computing environment*
R21	To be usable without being restricted to a commercial LCA software

Table 4.3 Definitions of Integrated, computing and engineering

	I	C	E
I	**Integrated**: To form, coordinate or blend into a functioning or unified whole	**Integrated Computation**: Combination of software and/or hardware tools in a functional unified system	**Integrated Engineering**: Combination of different processes and technologies. Implies increasing functionality in a reduced system size
C		**Computing**: To determine or calculate something by means of a computer	**Computational Engineering**: Also known as Computer-aided Engineering (CAE). It refers to all engineering tasks that are supported by computers
E			**Engineering**: Activities related to design, construction, manufacturing, recycling, and operation of complex products and their components and materials

of a specific performance indicator (e.g. plasticity in ME and global warming potential in LCE) followed by a top-down inverse engineering process considering the performance estimated as a chain of constraints. Furthermore, both research disciplines are characterized by an increasing trend on the development and application of computational models and virtual design support for product development. In the field of ME, the necessity of a holistic framework that consistently links multiscale engineering models motivated the emergence of the field Integrated Computational Materials Engineering (ICME)." This field has the purpose of accelerating the development processes of materials, improving the design process and its optimization, and integrating design and manufacturing (NAE 2008) (Fig. 4.3).

A formal definition of ICME is "the integration of materials information, captured in computational tools with engineering product performance analysis and manufacturing-process simulation. The main objective of ICME is to engineer the properties of products and their components based on the properties of its/their materials" (Schmitz and Prahl 2012). To this end, ICME pursues the integration of multiple disciplines through validated computational models in which structured information

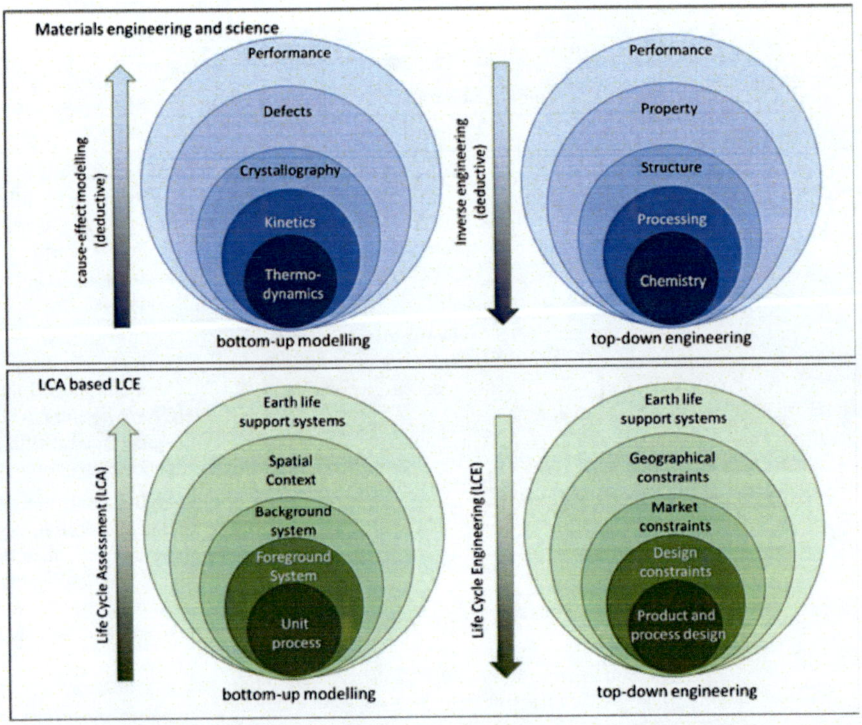

Fig. 4.3 Similarities between LCE and ME (own figure). The upper part was taken from the website of the joint Center for Computational Materials Design (CCMD) from Penn State University and Georgia Tech (www.ccmd.psu.edu). The lower part is inspired in Figs. 2.2, 2.4 and the Lyngby framework (Hauschild et al. 2017)

is transferred between different codes (TMS 2013; Horstemeyer and Sahay 2018). This link of information can be horizontal (i.e. codes linking sequential materials processes and their mechanical properties), vertical (i.e. codes linking the heterogeneous length-scale cause-effect relationships) or a hybrid (i.e. horizontal- and vertical codes are integrated) (Horstemeyer and Sahay 2018).

The common vision of ICME is to create an integrated platform that enables a "plug-and-play"-type combination of validated material models that are linked to the related manufacturing and product design models (Allison et al. 2006a; Schmitz and Prahl 2016). As stated by Allison and colleagues, "such a suite of tools in a robust, user-friendly computational environment would enable simultaneous optimization of manufacturing process and component design, materials selection, or rapid materials development with calculation of uncertainty metrics" (Allison et al. 2006a).

Current research efforts in this field are classified in mainly four domains being (Panchal et al. 2013; Taylor et al. 2018): (i) the integration of informatics tasks with the objective of ensuring a successful transferability of information within the stakeholders, (ii) development of scientific models that consolidates a solid theoretical understanding to extrapolate between experimental and historical data points, (iii) integration of empirical approaches from experiments, data fitting or machine learning and (iv) uncertainty quantification and design optimization in order to assess the quality and robustness of models.

While ICME is still a young approach (Allison et al. 2006a, b), it has already proven to bring industrial benefits.[2]

ICME rose from a similar set of obstacles and challenges that currently limit the application of LCE in product development. These are, in short, the lack of a computational methodology to analyze complex systems quickly, inexpensivly and comprehensively. The main advantages of ICME, drawn after different application cases, are that it can contribute towards: (i) reducing product development time and resources, (ii) innovations in material designs and therefore in product and process design, (iii) reducing the amount and scale of experiments, (iv) increasing product quality and performance and (v) towards developing new materials (Horstemeyer 2012). It is therefore reasonable to argue that LCE could profit from following a similar methodological evolution.

ICLCE Framework for traction batteries

The ICLCE concept builds upon the vision followed by the MSE community presented in Fig. 4.4. ICLCE is envisioned (Fig. 4.5) as a plug and play-like platform that enables coupling and freely exchanging multidisciplinary models in different

[2] Some examples of successful applications of the ICME approach in the aerospace, automotive and maritime industries are described in detailed in (TMS 2013). A project led by GE aviation resulted in a strong reduction of an expensive material used in super-alloys for engine turbines. In the automotive project, Ford reported to have achieved up to a 25% reduction of the product development process cost while reducing considerably the weight of the component produced. In another example, the application of the ICME approach led to the discovery of a new alloy after a period of two years. As stated in the report, such discovery projects last traditionally around six years.

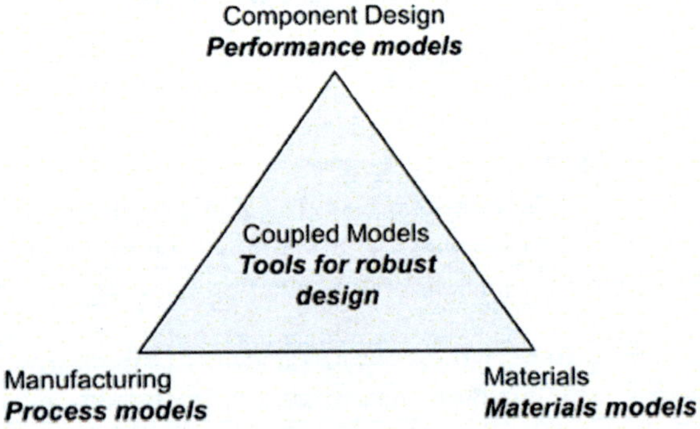

Fig. 4.4 ICME based on couple models. Based on Allison et al. (2006a)

scales for all life cycle stages of a product system. This flexible integration of multidisciplinary models enables a better consideration of the variability of parameters and expert knowledge, which consequently leads to a more robust support for design and engineering. The concept builds upon the reciprocity principle introduced by Cohen (1976). Cohen argued that it is conceivable that a dualism exists between the structure of a material and its properties, and between its properties and a particular performance indicator. This reciprocity principle has been interpreted by the ICME research community as a modelling paradigm in which a deductive cause-effect logic follows one modelling direction (i.e. from the structure of a material to its performance), whereas an inductive goal-means relation follows the opposite modelling direction (i.e. from performance to structure).

As reviewed in Chap. 2, current tools used to implement LCA present several weaknesses that hinder its application in industry to better support engineering activities. In the ICLCE, a bridge between environmental impacts and the different design parameters and system constraints is built through an inverse engineering modelling, as presented in Fig. 4.3. In this regard, it is possible to induce a specific parameter value based on its relation to a particular environmental impact indicator.

Four primary elements are essential in life cycle engineering: the system's design and development in terms of requirements, the product system (as presented in Fig. 4.1), the LCI and the environmental impact. While it is not easy to agree on how these elements are interconnected in reality, the LCA methodology has made it possible to linearly relate them. The ICLCE framework for traction batteries was developed considering the above, and is shown in Fig. 4.6. It is organized as a W-model describing the engineering process of complex product systems based on a continuous integrated computational life cycle engineering.

This representation of the modelling framework in ICLCE is inspired by the W-model for adaptronic systems introduced by Natterman and colleagues in (2013). A *W* model is derived from the modelling and development logic structured originally in the V-model for the development of software. The ICLCE W model is composed of two V models (left-side and right side V) which converge iteratively in a computational platform. A V-model is regularly used to describe the iterative development of complex systems while taking into account domain-specific design steps and whole

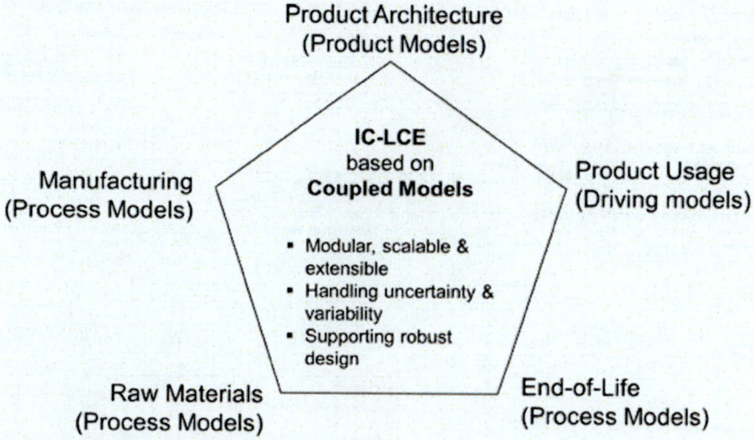

Fig. 4.5 Abstraction of the proposed approach based on Cerdas et al. (2018)

system integration. As argued in (Nattermann and Anderl 2013), the aim of the system design is to define the solution concept that will best describe the product under development. In the W-model for ICLCE, this in implemented in a similar manner, as is observed on the left side "V" in the Fig. 4.6. Here, process structures the modelling sequence following a hierarchic decomposition of the system into discipline specific components and their respective models. From Fig. 4.6, *vehicle engineering* is to be interpreted as a construct of multidisciplinary technical aspects, which range from the product's design and production issues such as geometry, material selection and processing parameters, to constraints in the background system and the regional context in which the traction battery is operated. A bottom-up model integration departing from the discipline-specific models takes place next.

As mentioned previously, both the development of traction batteries and their manufacturing processes are highly intertwined, so that measures for reducing their environmental impact within the design phase will strongly depend on modelling approaches that consider and link several disciplines. The ICLCE framework for traction batteries enables the integration of models from a relatively large set of disciplines involved in the development of these devices. This includes electrochemical models, material sciences models, detailed production process models for the production of the materials required, process models for the manufacturing of cells and battery packs, models to describe the energy consumption of EVs during the use phase, and recycling processes. For the case of EV, this is abstracted in the figure as a model decomposition from a top level system such as the *urban mobility system,* or the *BEV,* down to the components of which this system is composed (e.g. the battery system). These models are all linked to models in the background system, describing, for instance, the production and the composition of electricity mixes within a particular city's or region's grid, or urban transportation models in a region. Finally, ICLCE allows for the integration of models to research the influence a particular regional context might have within a specific life cycle stage of a product.

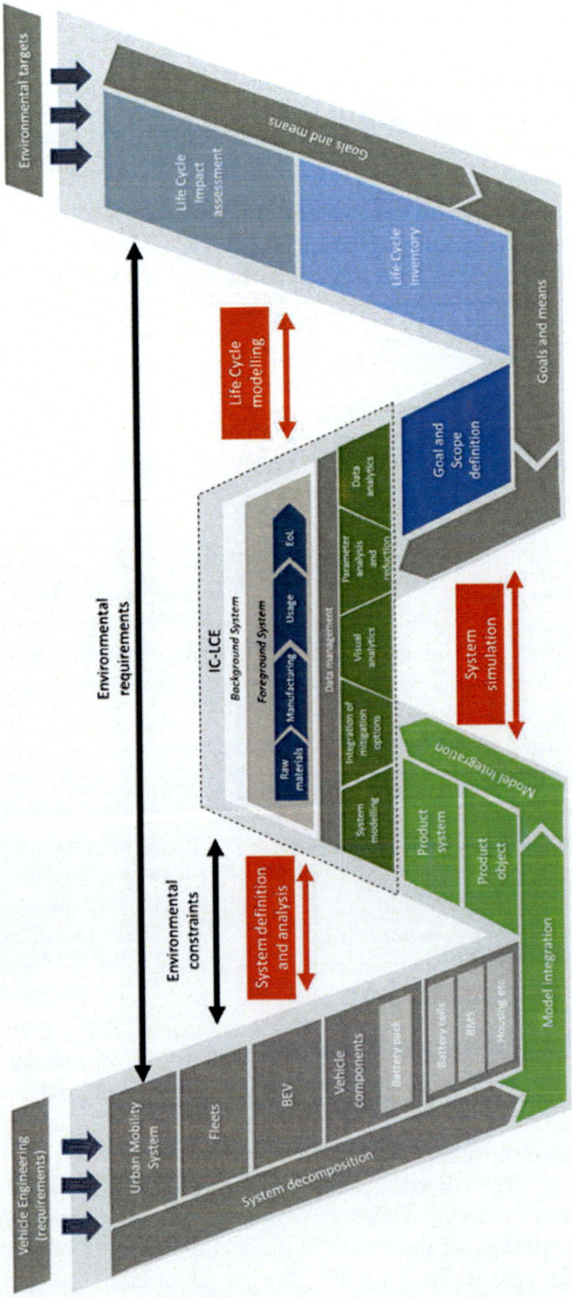

Fig. 4.6 *W*-model as modelling approach framework for ICLCE. Inspired by the W-model for adaptronic systems presented in (Nattermann and Anderl 2013)

All of the levels of models are linked through material/substance and energy flows following a cause-effect modelling which ends up with a final definition of a product system (see Fig. 4.6). In this regard, the entire system and the discipline-specific models are iteratively defined and analyzed.

During the product integration, a first interaction with the right-side V takes place. Here, an initial definition of the assessment goal and scope is done based on the system definition on the left-side V. With a defined product system, the ICLCE further quantifies linearly the elementary flows and proceeds to form structured LCIs, which are then characterized into environmental performance indicators following the traditional logic in LCA. This seamless execution of linked models allows a faster and more integral understanding of complex technical product systems such as, in this case, a battery system, without relying on rough simplifications that lead to losses of information and perspective. By following a series of property-performance relations through hierarchical sub-models (i.e. from a unit process to a life cycle environmental indicator), LCA is able to deduce the environmental impacts of a design.

While ICLCE is designed to model and calculate a large amount of LCA results for traction batteries in relatively short periods of time while considering foreground, background and spatial variabilities, it is also possible to integrate statistical and analytical algorithms to establish links through mathematical correlations of parameters with respect to extensive results and design parameters or system constraints. An inductive goal-driven modelling is followed for this purpose, i.e. from a results scatter down to single design parameters, such as the thickness of the cell's cathode coating, the energy mix of a manufacturing site, or the application that a battery system has been implemented for.

Reference architecture

While the development and coupling process of models can be time intensive, its execution and integration using computational approaches enhances its robustness, reliability and efficiency in terms of number of results per period of time obtained. In line with the objectives of this thesis, this section proposes a general reference architecture (RA) of an ICLCE model platform for the specific case of traction batteries.

An RA is defined as a formal description of a system to guide its implementation. It has the objective of managing the complexity, scope, size, updatability and extensibility of a software or system. RAs facilitate coping with trends such as increasing dynamics and the integration of emerging systems. In fields such as product and business development, and in systems engineering in general, RAs support shorter time to market, increased interoperability and systems reliability. A RA for the developed ICLCE system is presented in Fig. 4.7.

The objective of the RA for ICLCE is to guide the development of model-based tools with the purpose of estimating the life cycle impacts of these devices and integrating this information within product and process design and development processes. It is meant to be a template for a generalized solution of a system that is able to integrate a large number models in order to create a life cycle model for a traction battery system. The system is composed of a graphical user interface GUI layer, a data layer and five logic layers.

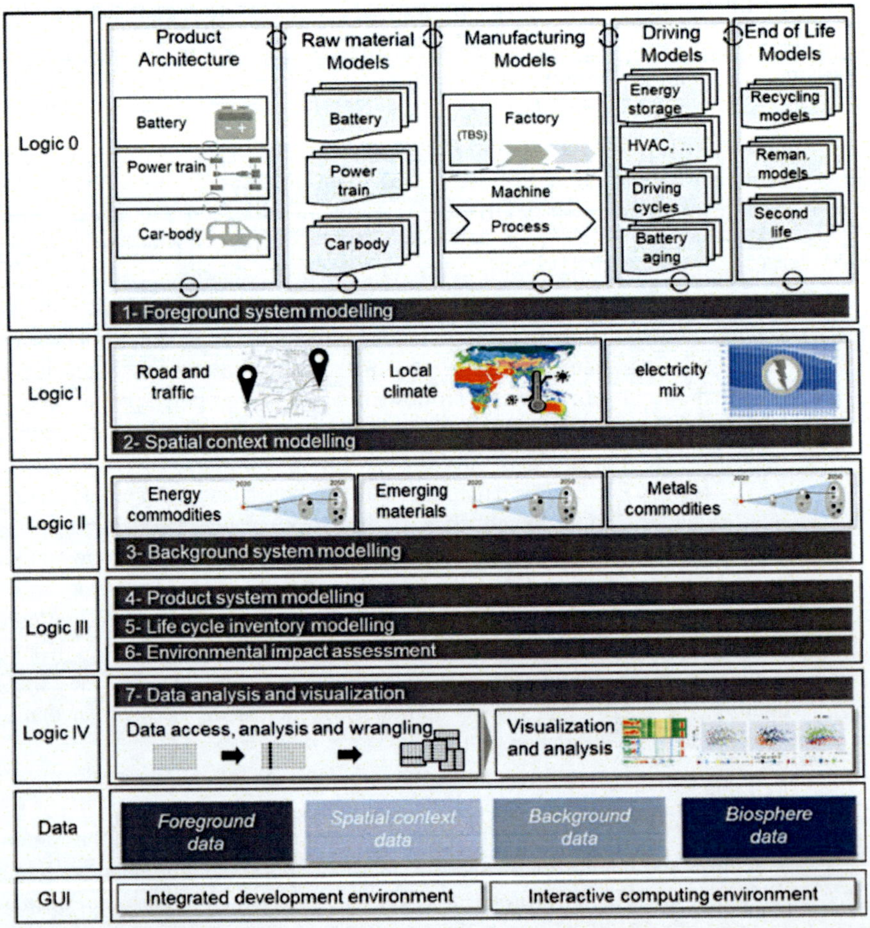

Fig. 4.7 Reference architecture for an ICLCE system for traction batteries. Further developed from Cerdas et al. (2018)

A logic layer is, in this context, a computational structure able to link different models through particular programmatic commands. The logic layers in the system contain libraries of theme specific models that can be called and integrated in the foreground or background system. Disregarding the research question being addressed, the seven milestones need to be completed in order to finish a modelling cycle. These modelling milestones are spread through the different logic layers, and need to be regarded as a guide during the development of life cycle models.

Logic 0 has the objective of modelling the foreground system (1) of a particular of a traction battery. These models can range from product architecture models to particular models depicting a certain unit process in the life cycle of a traction battery. Logic I has the objective of building the scenarios for the specific spatial context (2) in which a technology is to be operated. Logic II provide a modelling platform to integrate, manage and edit background models (3).

Logic III has the objective of building a product system based on the results from the coupled models in the foreground, spatial and background systems (4). Further, this logic has the task of developing a life cycle inventory (5) from this product system and calculating LCA results (6).

Logic IV has the main objective of arranging all the information created for one product system (i.e. user inputs, intermediate results, LCIs and LCIAs). It structures the results so as to generate visual insights useful in decision making processes (7). This logic receives arguments from the user (e.g. a command for a specific type of analysis such as uncertainty or sensitivity analyses). This logic contains a library of possible visualization analysis and methodologies that aim to represent not only the complexity of the product system but also to give insights that support decision making in life cycle engineering.

Logic IV provides, on the one side, access to the results stored in the databases. On the other side, as typically the ICLCE will be delivering product systems in the order of hundreds or even thousands, this logic is very useful in providing predefined data analytics and visual analytics support to ease the retrieval of information and knowledge that can be used in engineering decisions. While ICLCE has the potential to expedite the acquisition of the necessary data to support the life cycle engineering of complex product systems, the challenge within the decision making process increasingly shifts to the interpretation of the results. As with most complex systems, in life cycle engineering decision makers usually face the challenge of dealing with data from heterogeneous sources and quality. Moreover, the results in life cycle engineering are often conflicting and contrasting. As argued by Keim et al. (2005) having access to an overload of information not only means that significant efforts in terms of time and money are required to retrieve knowledge out of the data, but also that a great deal of relevant decision-making knowledge is not exploited to its fullest potential.

ICLCE is foreseen to support the data visualization and information processing methodology proposed by Keim et al. (2005). Visual analytics in this context is defined as a combination of "… automated analysis techniques with interactive visualizations for an effective understanding, reasoning and decision making on the basis of very large and complex data sets" (Keim et al. 2005). The motivation behind the visual analytics concept that the author presented was to "… exploit and use the hidden opportunities and knowledge resting in unexplored data sources" (Keim et al. 2005). The ICLCE approach provides access to every part of the data pipeline, enabling a higher degree of interaction. This aims to: (i) provide synthetized and understandable information and (ii) effectively communicate this information so that it leads to engineering actions.

The integration of the visual analytics concept in ICLCE was structured following the guiding questions proposed by Keim et al. (2005) and the discussion on visualization in LCA presented in (Cerdas et al. 2017): (i) what are decision-making procedures in life cycle engineering? (ii) what is relevant information in a decision-making process in life cycle engineering? and (iii) how to present results in life cycle engineering in a task-oriented way?

Figure 4.8 presents a framework for understanding the visualization requirements based on (i) and (ii). Visualization requirements were defined to fulfill both an assessment and an engineering procedure. Thus, the objective of this module is twofold: (i) modelling of all relevant information flows and data structure in such a way that it is possible to generate visualizations that feed and support an analytic discussion with discovered knowledge and (ii) proposing innovative approaches to visualize this information in the form of visualization methodologies. While assessment procedures have mainly a descriptive nature, the engineering procedures identified intend to disclose insights that are considered in a design or development process.

Regarding the possible assessment procedure, three actions are considered which are already common in LCA. First, the concept allows for a comparison against an alternative function (i.e. unit process A against unit process B, product system A against product system B, etc.). Second, the concept allows for the assessment of a given element's contribution to the environmental impact of the whole system (i.e. typical contribution analysis). Finally, the concept evaluates the proportion or significance of a given element with regard to a given context (i.e. normalization).

Regarding the engineering support procedures, three relevant aspects were identified in addition to a comparison of alternatives. First is to understand the sensitivity of specific parameters in terms of the whole. This analysis goes beyond the common sensitivity analysis in LCA, in which the weight of an assumption about the final LCIA results is analyzed. Here, the sensitivity of a parameter is tested by looking at its influence on other parts of the life cycle model, describing the product system. This is useful, for example in the instance of developing parameterized models to execute lighter and faster analyses, or to be integrated in optimization algorithms. Second, the analysis of tradeoffs throughout the hierarchic structure of the product system and throughout the elements composing the product system is a further relevant procedure. This is important as it enables a higher transparency that leads to the easy detection of parameter correlations and conflicting goals. Third, by bringing the previous two procedures together, it is possible to define design spaces or constraints at the different levels previously defined.

In ICLCE, four levels of analysis and action (i.e. levels in which an engineering decision can be made) are defined: unit process, product object, life cycle stage and product system life cycle.

The third part of the visualization framework in ICLCE regards the method of communicating the information required for each of the procedures. While there are multiple representation options to show the same information, this part of the framework aims to guide the selection of the chart needed. In this regard, the visualization framework proposes a matrix relating the source of potential variability (technological, temporal and geographical) and the type of procedure supporting decision making in life cycle engineering.

The data layer contains all of the data that is fed into the different logic layers. This includes technical data for the models in the foreground system such as values

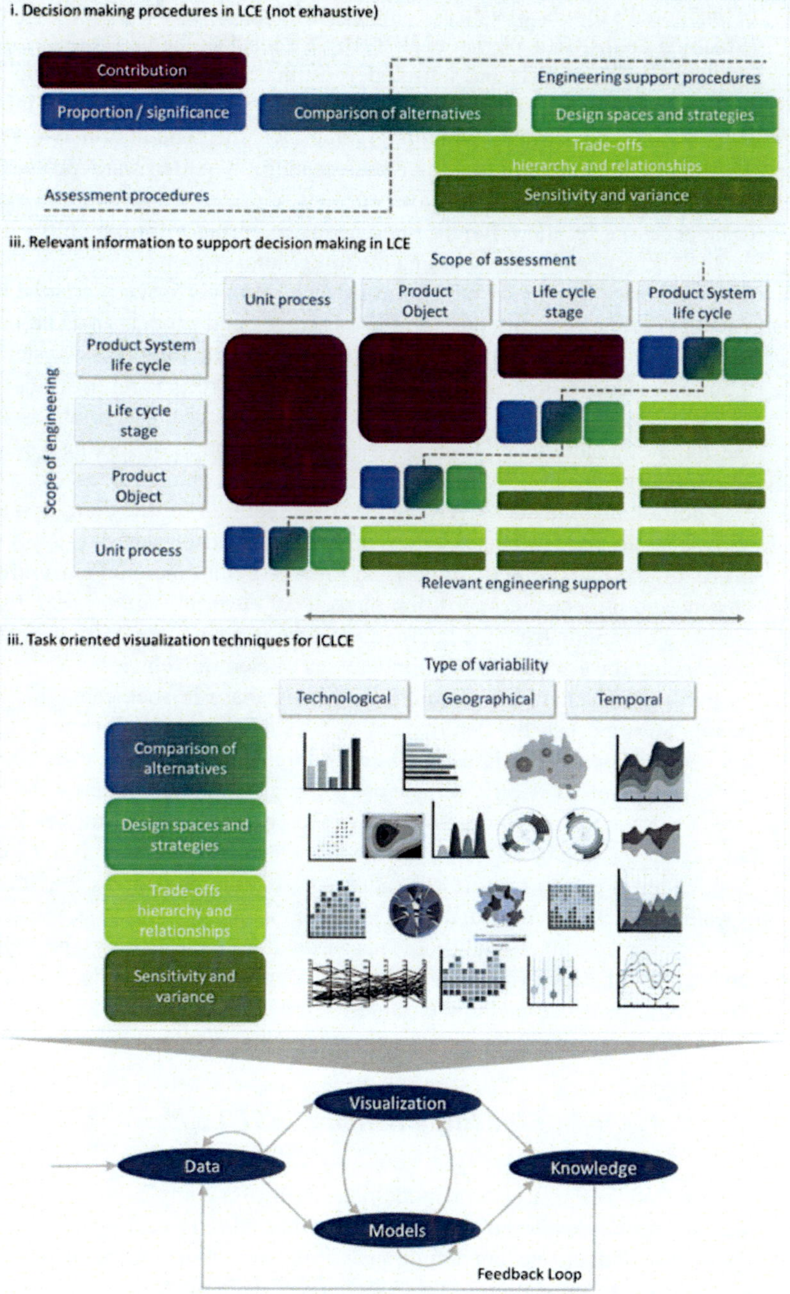

Fig. 4.8 Visualization framework in ICLCE. The bottom part of the figure is based on Keim et al. (2005)

of the model's inputs, as well as commercial background databases such as ecoinvent (Steubing et al. 2016; Wernet et al. 2016), for instance. It also contains intermediate results from the models and sub-models estimated at the different logic layers.

The data layer also contains relevant regional information, such as infrastructural and transportation aspects like topographies, weather, and road-networks, as well as socioeconomic indicators and data such as mobility patterns and profiles, city sizes, and population densities. This layer consists of organized embedded data structured to be accessed by the different logic layers requiring information for its operation cycle. It is responsible for sourcing the foreground system (i.e. processes and models of the product system for which the LCA is being performed) and the background system (i.e. the system consisting of processes and models that are not the main focus of the LCA but that have strong influence on the foreground system) with data inputs. The foreground data embedded in the system refer to repositories of discipline specific models (i.e. physical, data-based, empiric, etc.), describing aspects such as the aging of a battery cell, average energy consumption by mixing a batch of active slurry, energy density of a particular cell chemistry, vehicle sizes, etc.

Additionally, the data layer includes data required for the definition of scenarios in the background system such as: (i) regional aspects of material supply chains (e.g. origin of and market mix composition for a particular material), (ii) information on current and projected composition of regional electricity mixes (e.g. from national statistics or reports from the international energy agency) and (iii) geographical data influencing the environmental implications of electric vehicles.

Further data that can be stored in this database is socioeconomic information having an impact in the deployment of EVs (e.g. statistics on car ownership, new EV registration and political incentives). Inter-individual data affecting the energy consumption of an electric vehicle per trip such as driving patterns and behavior for specific driver groups, or emissions and average commuting times and distances for a specific city, can be as well integrated. The developed system is not foreseen to be applied as an end-user tool. It is rather a platform that enables the development of specific tools. In this regard, it is not possible to interact with the system as is typically done with a computational tool. This means that there is no graphic interface that enables the input of values or selections. Depending on the desired application, a GUI might be developed in the future for the end user of the tool.

4.3.2 General Modelling Scheme in ICLCE

Velten (2008) offers a general modelling scheme for complex systems which has been adapted for the development of integrated life cycle engineering models. The scheme can be divided into system definition and analysis, life cycle model development and simulation. A description of these sections is given next.

System definition and analysis

The objective in this stage is to define the research question and the problem to be solved. Moreover, a full definition of the product system needs to be performed which, in a general sense, implies the identification of all of the system entities that will be considered and coupled in the life cycle model.

In LCA, the scope of the life cycle model that is being developed needs to be defined before starting with the modelling phases. This preparatory step aims to guide the model integration process and is generally not tool-supported. Typically, this phase includes defining the research question which, broadly speaking, is inherently related to the functional unit. The functional unit in a product system is the reference unit of functionality being provided by a product system, and therefore by the life cycle model. In the present modelling approach, this step is simplified due to the bidirectional coupling of the models that allows tracking the physical properties of the PS as described further in this chapter. This means, in the case of a traction battery for instance, that the final results from the LCIA can be normalized to usual specifications such as specific energy density, mass, and energy capacity, to mention some. Moreover, since all of these physical properties are being estimated, tracked and linked to the product system, enhanced FUs can be defined during the data and visual analytics modelling phase (e.g. including battery life time or even safety properties in the definition of the functionality of the PS).

The scope definition step additionally includes the definition of the boundaries of the product system that will limit the developed life cycle model as represented in Fig. 4.9.

Within the ICLCE approach, this implies a full definition and development of models in the different systems composing a product system. The definition of foreground system boundaries and models refers to the group of unit processes for which the LCA is being carried out, namely the processes which can be influenced by the decision maker. In ICLCE, this step includes additionally the identification of the models that are required to be developed and further coupled. This implies a definition of interfaces between the models that will be integrated and an analysis of first-

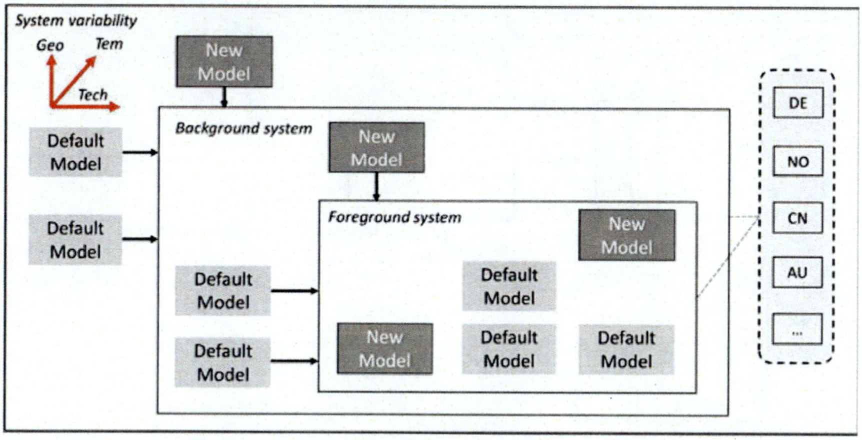

Fig. 4.9 Scope definition and model integration in ICLCE

and second-degree interactions between input and out parameters in downstream models. The background system refers to selecting the background models that will be edited and linked to the models in the foreground system. In this step, all of the models in the background system that need to be developed and coupled should be identified.

Finally, defining the geographical system boundaries and identifying relevant required models in the spatial context system is required in ICLCE. The objective in this step is to identify the locations in which a particular unit process takes place and define the proper linkage to the models in the background system. Additionally, models depicting influences of spatial factors to the operation of the processes in the foreground and background system might be identified and coupled in the life cycle model.

Life cycle model development

The objective in this step is the coupling of the models in an integrated life cycle model to answer the research questions defined. Following the most common epistemic definition, a model in ICLCE is defined as a simplified description of a real system that can be used to answer questions about the system. A model has, in other words, the objective of representing a physical relationship or transformation process of a set of given inputs. This is described in Fig. 4.10, in which a model works as a receiver of several types of inputs and as a generator of outputs. Inputs in this regard are: (i) data, (ii) variables and (iii) objects (e.g. models, dataset, etc.) whereas outputs are objects containing information required by the models downstream.

The investigation of the life cycle impacts of complex systems through the application of an ICLCE approach is represented schematically in Fig. 4.11 in the form of an input–output system. Important modelling properties are essential in this regard

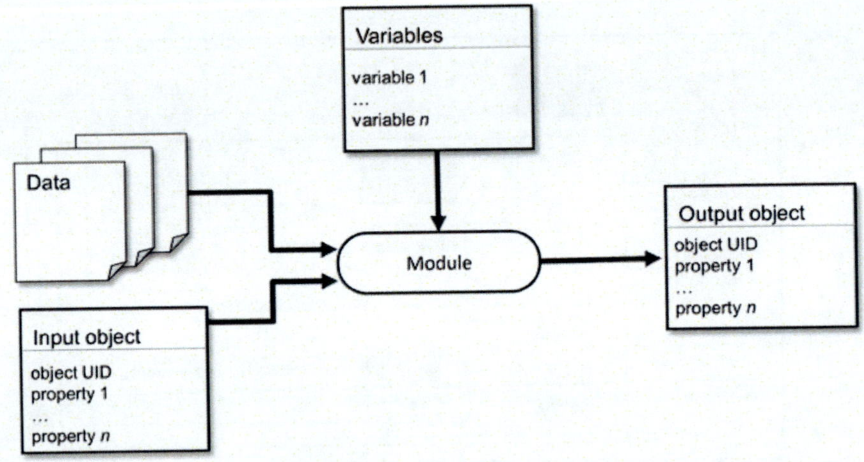

Fig. 4.10 Abstraction of a model in ICLCE

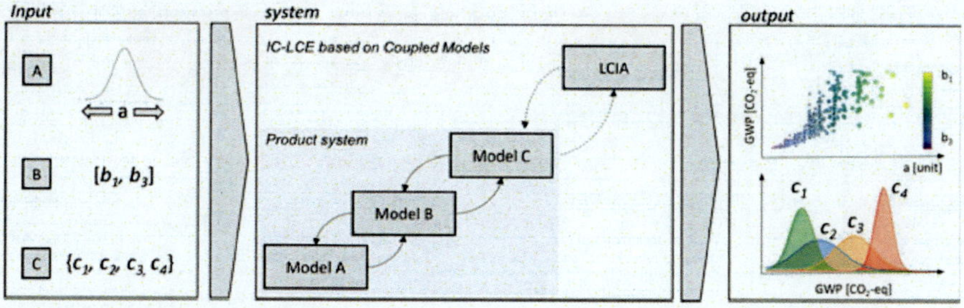

Fig. 4.11 Overall modelling approach as an input–output system

as implicit in the figure. First, the ICLCE is able to integrate domain specific models, which means that the models are freely exchanged and coupled respecting determined interfaces. This is important, as the concept should enable researchers to couple discipline-specific models in a larger system model in order to examine the effects of the technology being developed from a life cycle perspective. The result of this discipline specific model interacts with other models and consequently affects the virtual product system that the ICLCE generates. This means, for instance, that if a new processing technology is being developed to mix and coat slurry pastes to produce electrodes, a model representing this new process should be able to be coupled to an LCA model without the need of developing a full LCA for that specific process.

Rather, coupling the model to an ICLCE model would make it possible to evaluate that new process within many different scenarios, and to give useful feedback that can be used to optimize it. Seven modelling milestones have been identified as necessary in order to complete the modelling cycle. Figure 4.12 describes the sequence of phases in which a life cycle model is developed and executed for one product system. As in a traditional LCA, the first step is the definition of the scope of the model, in which the boundaries of the study are defined and the models are coupled. In the next phase, an entity called product object (PO) is built through the interaction of the models in the FS. Next, if necessary, the models in both the BS and the SC are run and prepared to be programmatically linked to the PO in the following modelling phase to construct the product system (PS). Once the product system has been properly created its data structure is transformed so it matches common data structures of environmental LCIs and it is uploaded onto a database of activities.

The activity is next called by an LCA script, which reads all the technosphere and biosphere exchanges, as well as the links to commercial or newly defined background systems and estimates the environmental impact for each impact category desired. The results are then linked to the PS, integrated in its data structure, and stored in a database of evaluated product systems. In the final modelling phase, these evaluated product systems are read from the database and the information is structured to facilitate the development of charts.

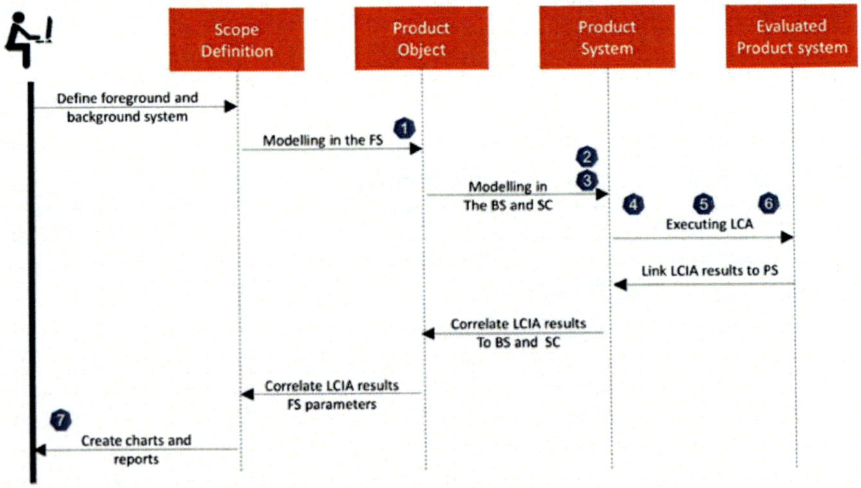

Fig. 4.12 Sequence diagram of the modelling steps. 1–7

Simulation

Simulation is understood from its original meaning where a simulation cycle is essentially the application of the developed model with the objective of understanding and interpreting the results, in terms of which insights are obtained to solve the defined problem. This understanding of simulation in this context should not be confused with more modern definitions, in which the behavior of a system over time is the main focus of the model. The nature of these models to be integrated in ICLCE, regarding the type of inputs and outputs, may differ among the models integrated in the coupling platform. So, for instance, the values of the input parameters might have the form of a given statistical distribution, a range or a set of discrete values. Nevertheless, the modelling logic enables a stochastic combination of these parameters (e.g. within a Monte Carlo simulation algorithm) which leads to a large set of results, which can then be used to develop visual analytics tools. Further, every relevant information flow is traceable throughout the modelling process, which in turn enables the identification of first and second degree effects on the final results through the application of variance based sensitivity analysis process, such as Sobol as developed by Saltelli (Liu 2008; Saltelli et al. 2010) and exemplified for LCA by Groen and colleagues in (2017). Additionally, this bidirectional traceability eases the development of visual aids to understand, for instance, the effect of the input values in each of the models, their correlation (if any) with other parameters and the final results, the effect of specific group of values through clustering, and the influence of specific discrete values to the final results.

In the following sections, the aforementioned modelling phases are described. First, the FS modelling approach for traction batteries is described. Further, modelling approaches in the BS and the SC are discussed, and finally the modelling approach for the linking of information to generate a product system is presented.

4.3.3 Foreground System Modelling

The scope of the modelling in this logic layer is the coupling of domain specific models within the battery value chain, which leads to the definition of a *product object* (PO). Typically, these models were identified and developed in the system definition and analysis step. For the case of traction batteries, the reference architecture classifies these into five different groups which are inline to the groups represented in Fig. 4.7: (i) product architecture, (ii) raw materials, (iii) manufacturing, (iv) usage and (v) recycling. These groups serve as the basis to structure the modelling sequences, as in the proposed modelling logic in the foreground system is shown in the activity diagram in Fig. 4.13.

The figure shows the way the different modelling sequences in the foreground system are structured and coordinated. In a first modelling step, the set of models in product architecture for a given application are coupled and executed. The modelling step *product architecture* has the objective of defining the qualitative aspects of the product, such as its product geometry and the type of materials of which it is composed, and quantitative properties such as its size, volume, mass and material distribution. This step creates a product object that is stored in the database *foreground system.*

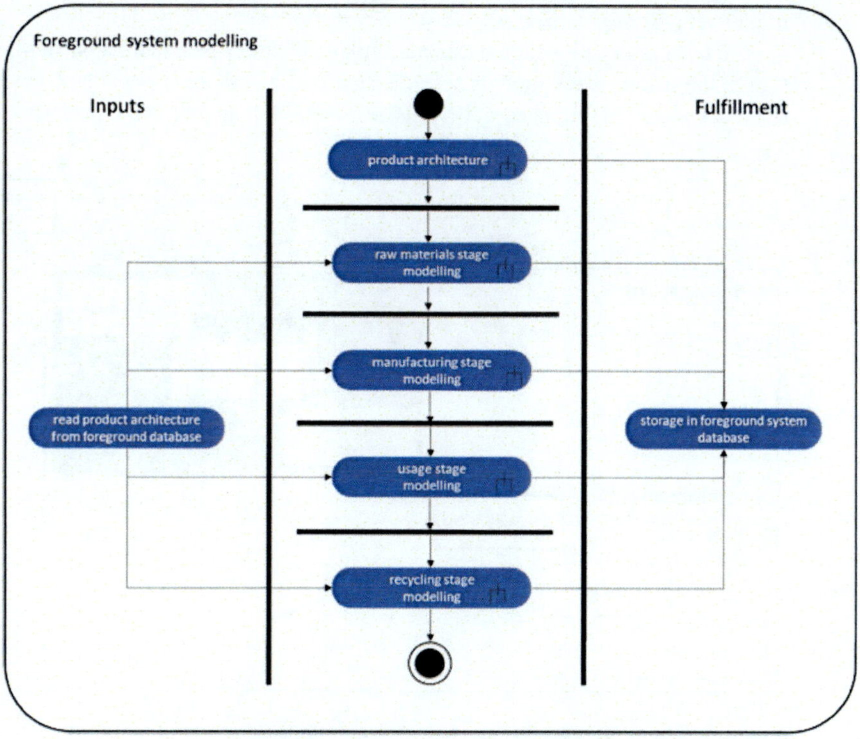

Fig. 4.13 Activity diagram of the modelling sequence in the foreground system

In subsequent steps the life cycle stages are modelled. In each of the phases, the information regarding the product object is retrieved from the database *foreground system* in order to consider relevant inputs to the models integrated in these modelling phases. Each of the steps updates and stores the product object in the database.

After the last step, a product object is stored in the database which already embodies all relevant values of parameters defined in each of the phases.

For the case of traction batteries, ICLCE structures the model coupling platform as shown in Fig. 4.14. Two main level of models are in this regard predefined, the level traction battery (1) and an application level (2) (e.g. EV). The level *application* imposes constraints to the level traction battery such as power requirements, energy requirements, and mass and volume limits, to mention a few. The level traction battery is structured into battery cells (1.1) and battery pack (1.2). *Battery cells* is constrained by battery pack through the available mass and volume for cells in the modules. Nevertheless, the geometry properties of the cells are independent from the battery pack.

Notice that the execution of the models in battery cells and the models belonging to the application might be executed independent from each other. Once the top-down requirements in the application level and the bottom-up cell properties in the battery cells definition are calculated, the level traction battery can be executed.

This means, in other words, that the architecture of the cell can run independently from the architecture of the traction battery. The logic inside the product architecture model set is shown in Fig. 4.15.

The *battery cells* level contains a number of battery cells with a particular definition to the battery pack models set. It is structured in three sub-levels: (i) electrodes, (ii) stacks and (iii) finished cell. This set of models gather the inputs from battery

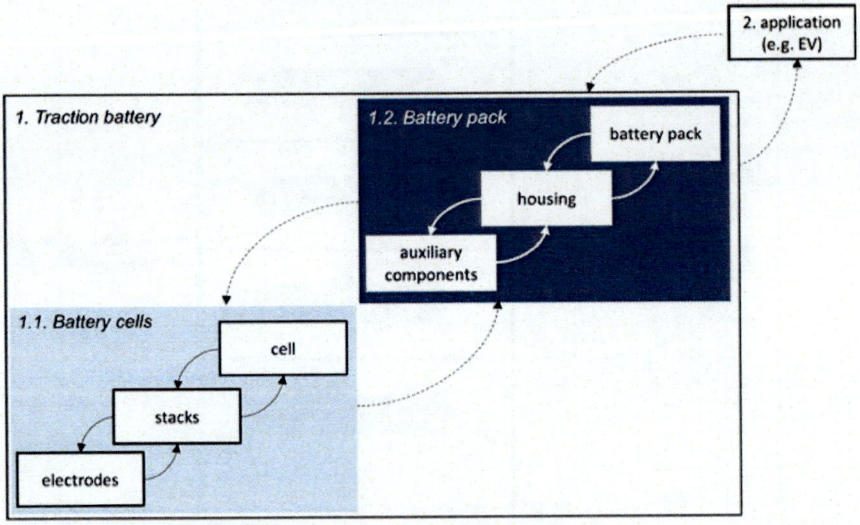

Fig. 4.14 Proposed product architecture model coupling structure

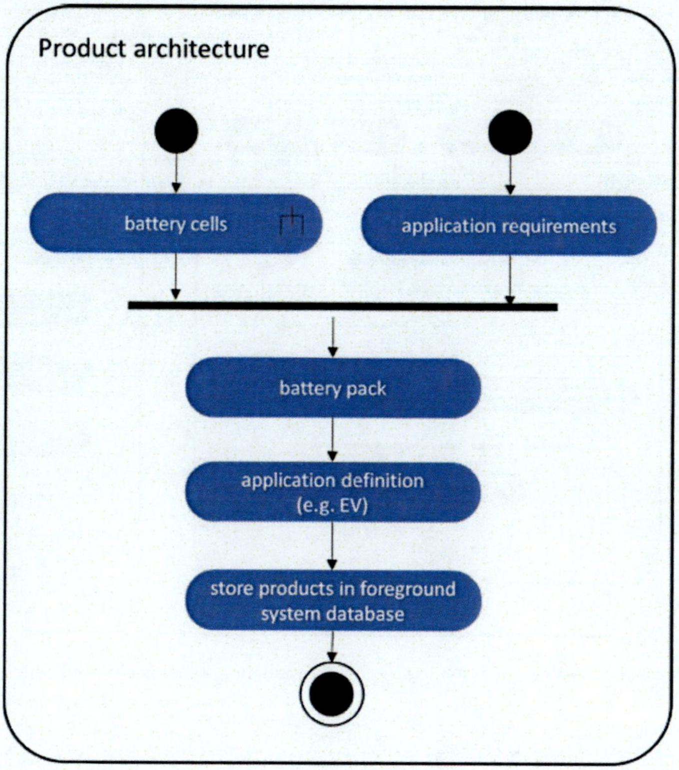

Fig. 4.15 Proposed activity diagram of the modelling sequence at the product architecture level

cells and application requirements and configures the traction battery required by the application vehicle taking into consideration the geometry and properties of the cells modelled. The logic to configure the battery cells is described in the activity diagram in Fig. 4.16.

The modelling in this phase starts with a set of models which aim to define the physical properties of the electrodes and store this information in a database. This is done while considering the validated experimental and/or theoretical properties of the active materials which are being modelled. In a further step, the models in stack definition gather the information of the electrodes from the database and configure the stacks which are stored in a database as well. The final step in this level is the definition of the final assembled cell. First, the geometrical specifications of the cells are defined (e.g. cell geometry, type, housing, etc.) and used to estimate the number of stacks required in the cell. The information regarding the stacks is read from the database and consolidated in an object called *assembled cell* which is afterwards stored in another compartment of the foreground system database.

Next, the remaining groups of models defined to represent the life cycle of the traction battery are sequentially executed, as shown in the activity diagram in

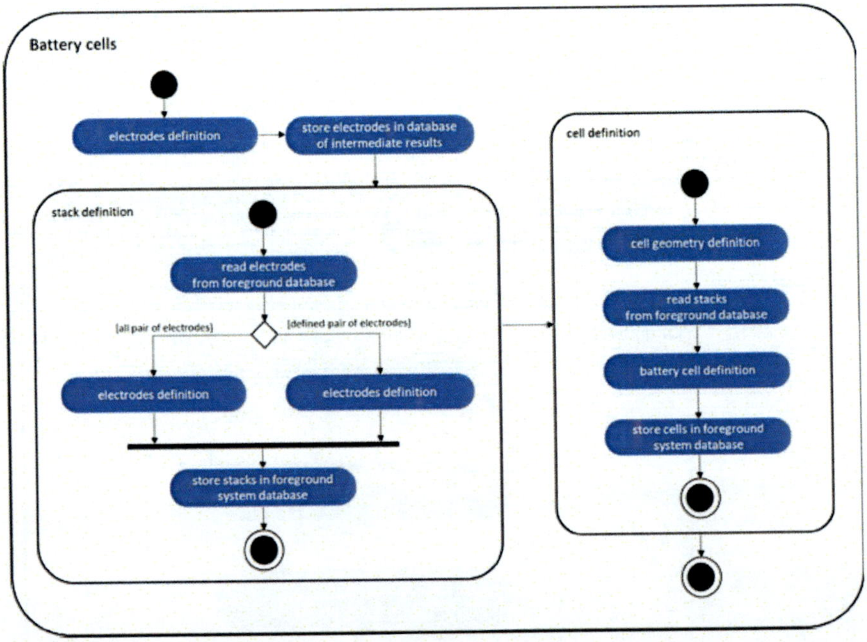

Fig. 4.16 Proposed activity diagram of the modelling sequence for the definition of battery cells

Fig. 4.13. The modelling step *raw materials production* reads the specification from the database foreground system and extracts the required information to start the modelling sequence. Two aspects here are important, namely the type and name of the materials contained in the battery system, and its quantity. The raw material phase separates these into active and inactive materials, active materials being the ones applied to produce the slurry mixture coated onto the current collectors. The production of the active materials in the ICLCE approach is structured into production of precursors and production of cathode and anode materials. Similarly, the manufacturing and end-of-life modules modelling steps retrieve all the specifications concerning the architecture of the traction battery from the database. The manufacturing module registers all relevant product parameters affecting the energy and/or material consumption of a certain unit process (e.g. electrode thickness influences coating and drying processes, coating porosity influences the quantity of electrolyte, and cell energy capacity influences the aging and forming processes).

The models for the manufacturing processes are divided into electrode manufacturing, cell assembly, cell aging and forming which are the stages within the manufacturing processes of a battery cell that are most influenced by the properties of the product. The EoL module retrieves all relevant properties of the traction battery (e.g. mass distribution for instance among active and inactive materials, cathode materials, etc.) and uses this information to estimate the material and energy flows of specific recycling paths through, for instance, the application of process-based models for disassembly, mechanical separation, hydrometallurgical and pyro metallurgical material separation. Foreground system models for the usage stage of the application vehicle will include at least a module for the estimation of the energy consumption per driven kilometer and range, and a module to estimate battery state of health, or the capacity losses of the traction battery so to be able to estimate its life time. Ideally, the models integrated in this layer and representing the material production, manufacturing, usage and EoL stages will consider all relevant product properties estimated during the product architecture modelling phase.

4.3.4 Background System Modelling

The models in the module background system are linked to the foreground system (see Fig. 4.17). While these background system processes usually represent a high share of the environmental impacts associated with the life cycle of a traction battery system, so far the possibility of increasing their transparency and adjustability has been limited. In this logic of the ICLCE, it is possible to access these models and edit linkages or compositions in the datasets based on particular models created for this purpose. The overall modelling approach pursued in this phase is represented in Fig. 4.17. The objective as seen is to create a new or edited dataset that will be stored in the background system database. In this way, it is possible to create background system datasets based on scenarios which will ultimately give useful insights regarding the environmental impact of a traction battery under variable circumstances in the background system due to, for instance, technology efficiency, electricity mixes, or transportation mixes.

The general process starts by reading a given dataset from the commercial database.

The objective is to identify and, if required, edit the datasets that will be linked to the technosphere flows from the foreground system. The first step is an automatic

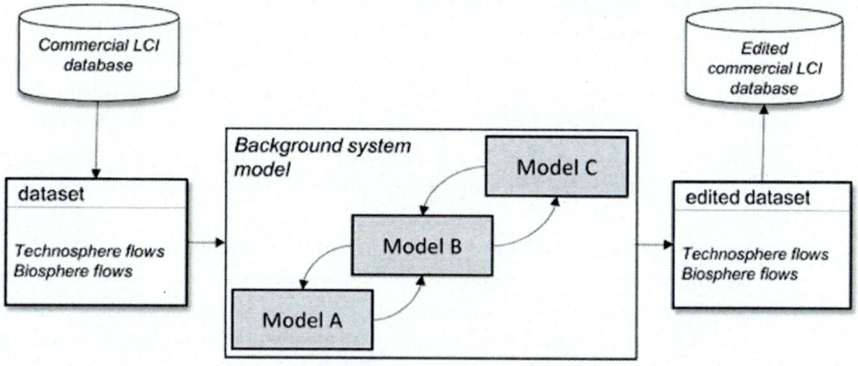

Fig. 4.17 General background system modelling approach

retrieval of the technosphere exchanges resulting from the models in the foreground system layer. In order to do this, the system reads through all the datasets created in the FS modeling layer, extracting the name and type (biosphere or technosphere) given to the exchange. Technosphere flows are then identified and their names are broken into keywords. These are then given to a programmed search engine that looks up for the most appropriate dataset in a given commercial database. The dataset is extracted and converted into a python dictionary, allowing to extract its unique identifier code which will be used to link the created exchange of the foreground dataset with the commercial database.

The commercial dataset might need to be edited. This might be the case, for instance, if the purpose of the analysis is to evaluate energy mixes scenarios linked to a given process in a life cycle stage. For the particular case of traction batteries, the following cases were identified in which the background system may be *model-based* edited.

First, editing may be required in order to link spatially explicit electricity mixes to relevant and sensitive background processes, such as the production of important raw materials like aluminum, nickel, and copper, among others. Additionally, it may be necessary to create or edit the electricity mixes for a particular place and time: this is relevant, as the time variation of the electricity's grid changes during the year due to the fluctuation on the share of renewable energies. The linking process can be executed without the need to create a new data set for an electricity mix. This is the case, for example, when the dataset of a process defined for a particular region (and therefore linked to the electricity mix of that region) is to be evaluated while running in a different place. While the implementation has been done to adjust datasets to model specific electricity mixes, the same modeling approach can be followed for the purpose of assessing changes in efficiency of the technologies represented in the background processes. This has been implemented by Cox and colleagues in the tool called *Wurst,* which enables relating the linear relationship between inputs and outputs of a given dataset with a technology efficiency factor. One example of the application of the tool is presented in (Vandepaer et al. 2018), where the efficiency of background transportation processes and electricity production technologies is changed in order to estimate the relevance of the impact driven exclusively by the background processes. Once this dataset is created and stored, it can be linked to a commercial dataset (for instance, production of aluminum or a recycling process). After a dataset has been identified and extracted, its specification in terms of inputs and outputs is extracted and structured as a python dictionary. The specification of this dictionary might be edited by exchanging the default electricity exchanges against the new dataset created one step before, thus linking the dataset to other commercial datasets.

Second, editing might be required in order to create scenario based electricity mixes and to explore the effect of a technology in future mobility scenarios. For instance, to explore the effect of energy transition scenarios for the manufacturing stage of battery cells or a vehicles usage stage. A new data set containing a specific electricity mix can be created by inputting its normalized distribution linked to the name and source. An empty data set is then automatically created and filled with the linked exchanges from the commercial database (electricity, high voltage, hydro, river), its normalized value as the input and its unique identifier. The output of the dataset is then set as 1 kWh of electricity and the dataset created is further tagged and stored in the database.

Finally, editing may be necessary in order to explore the change in efficiency of technologies in the background system based on the model. For instance, to explore the effect of the electrification of the transportation sector in the materials production value chain.

4.3.5 Spatial Context Modelling

In the layer dedicated to modelling the spatial contexts in which a unit process takes place, several factors leading to an increased variability in the modelling of the life cycle inventory of a product system are considered. These range from a grid's electricity mix being provided, to a manufacturing site, to the estimation of daily temperature profiles of a specific city, to the driving profiles of a region. This layer aims to create, read or write an object "location" that contains all the information that is necessary for the downstream model, as seen in Fig. 4.18. This object will be stored in a database as a set of linking rules. Once a user or a model calls the database to make a link to a specific location, the object including the information for this location is automatically searched for in the database, and linked to the process or activity being modelled. The spatial modelling is completed after five steps. The first step is the interpretation of the query requested by the user or by any of the models in the other layers (see Fig. 4.18). The query is submitted by specifying the location where a particular activity takes place. The specification can be done by inputting the codes defined by the ISO 1366 standard, which gives codes to represent the names of countries and their subdivisions. For instance, "DE" or "DEU" for Germany, or "AR" or "ARG" for Argentina. If the objective of the active spatial modelling routine is, for instance, to provide information for the route being followed in the driving phase, the query can be done with the city's name, a specific address or a set of coordinates. Once the query has been submitted, the following module extracts the necessary spatial information for the location in question. Once the required information has been recovered through the API links, the next step in the spatial modelling processes sequence starts. This step is called *linking in place* and its modelling is represented in Fig. 4.18. Linking in place is composed of an interpretation module, a module to search in the databases containing spatial information, four modules to link spatial information documented as a set of coordinates or polygons describing a given space, and a consolidation module.

To do this, different geocoding services providers can be used. The output is a GeoJSON (Butler et al. 2016) data file, including a description of the electricity mix being supplied to this specific location, a temperature profile, a driving profile, a list of available spatially differentiated characterization factors, as well as other important geo properties, such as shapes of polygons. GeoJSON is an open data

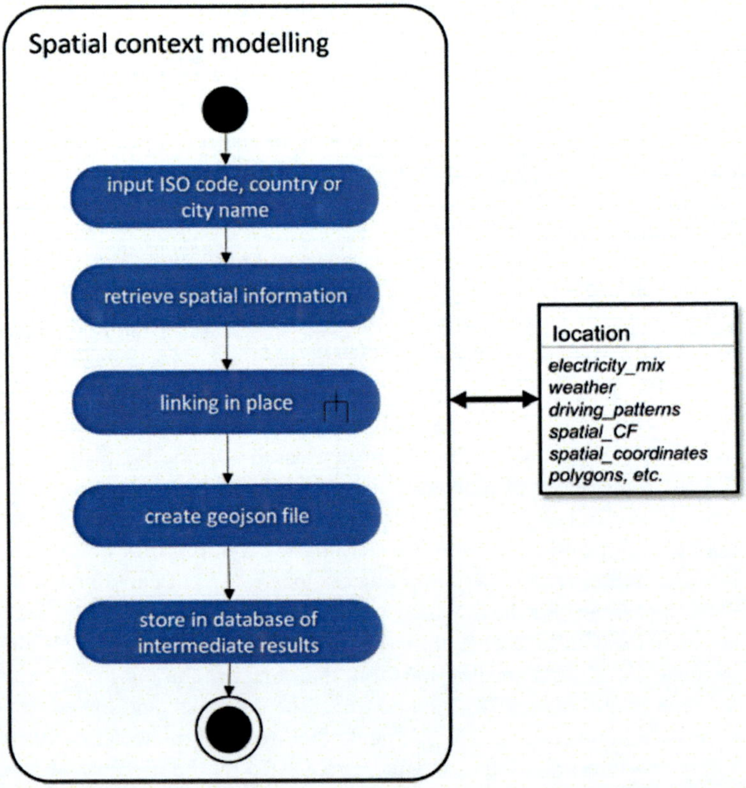

Fig. 4.18 Activity diagram of the spatial context layer

interchange format based on the JavaScript object notation. It was design to represent geographical features linked to other non-geographical properties.

The format entail features such as polygons (useful to represent regions, countries or cities), points (useful to represent a position) and lines and lines strings (useful to represent streets, routes, etc.) (Butler et al. 2016). This notation is useful within the ICLCE framework due to two main reasons: (i) it enables for the storage of information relevant to the modelling of life cycle inventories so that it is possible to consider the spatial factors influencing the performance of an activity and (ii) it allows for tracking where the processes take place (emissions are emitted, energy is consumed, waste is produced, etc.), which is later important when developing visualization methodologies to support decision making.

4.3.6 Product System Modelling and Assessment

Product object concept

A PO is a conceptual entity of a product system that is built after the foreground system modelling step and is, in its basic form, unlinked to the background system. In practice, a PO is a data structure that stores all pertinent information that is processed and generated by the different discipline-specific models integrated in logic 0, i.e. a PO contains all values of the parameters input to a model and its results. These results generally include all materials and energy flows emerging from each of the life cycle stages that are intended to be linked later to the background system and the spatial context in order to build the product system. These flows are classified into technosphere and biosphere exchanges and product flows. A product flow can be seen as the functional unit of a product object. This information is structured, linked and stored in a database in the foreground system layer and provides adequate encapsulation for other POs, attributes and life cycle relationship models as represented in Fig. 4.19.

The figure shows a representation of the PO *battery cell* which considers other POs, such as electrode stacks and cell packaging, as well as a relationship link to the four conceptual life cycle stages of a cell: *raw materials, manufacturing, usage* and *recycling*. The structure of a PO provides the basis of the modelling coupling methodology in ICLCE. In the ICLCE there is no semantic limitation regarding the composition of a PO in terms of the ontological relationships that define it. In this

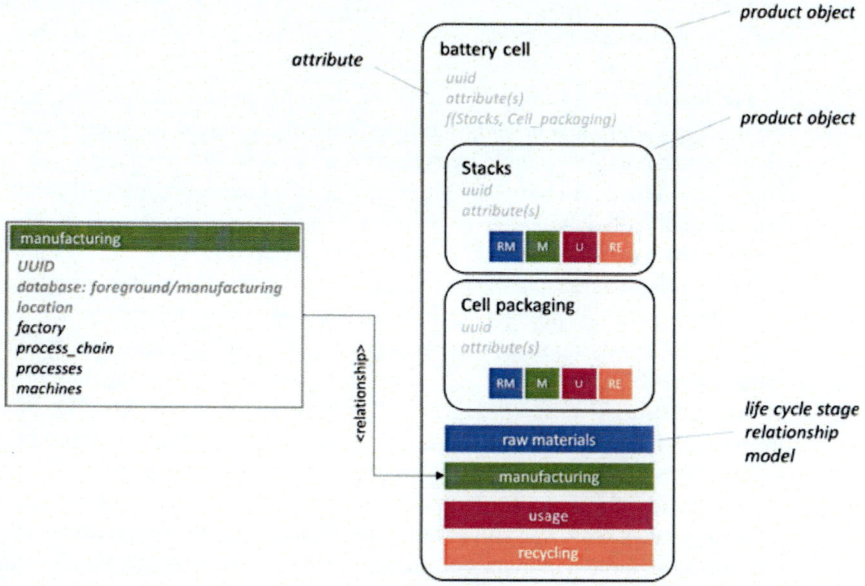

Fig. 4.19 Exemplary definition of product object—PO

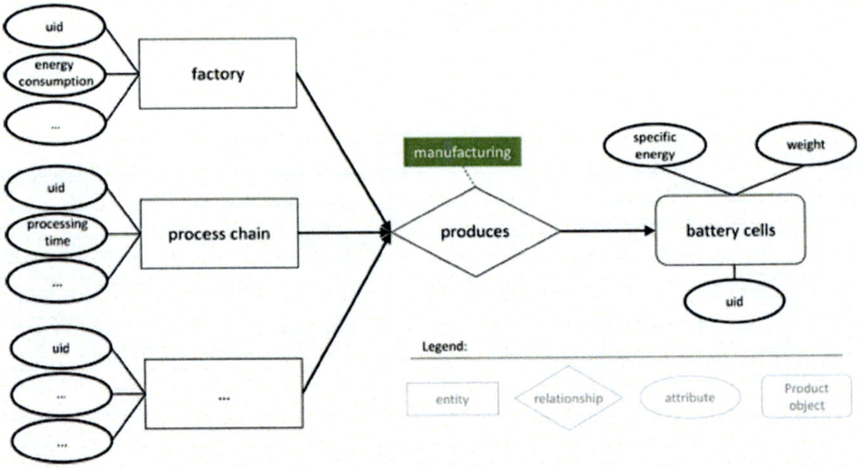

Fig. 4.20 Entity—attributes—relationship model in a product object

regard, a PO can be seen as a stand alone component or part of a larger PO (e.g. a battery cell). or as a complex product hierarchically structured and composed of other (sub) POs (e.g. an EV). A life cycle relationship model is an abstract link between the PO attributes and a physical or conceptual life cycle stage of the PO. Generally speaking, a relationship expresses the way in which two or more entities are related to each other. This is better illustrated with the example in Fig. 4.20. In this example, several entities such as *factory* or *process chain* are related to the PO through a life cycle stage relationship model *manufacturing*. The entities are in turn characterized by attributes.

For instance, the energy consumption and the processing time are attributes of the entities *factory* and *process chain* respectively, whereas the entity *battery cells* is characterized in this example by the attributes *weight* and *specific energy*. Additionally, since these entities are capable of independent existence, each of the entities are programmatically labelled given a universally unique identifier (uuid). This enables the information to be stored, combined and queried for.

In a similar manner, the PO is structured following a holonic hierarchy, i.e. starting from the top at the product level down through the components and parts until reaching the most important materials at the bottom. This is represented in Fig. 4.21. Here, the path from electric vehicle down to the active materials is represented. It is structured in six different levels, each containing a set of POs. The POs at each level are capable of independent coexistence and are related to each other through the relationships in the PO located on the level directly above them.

To illustrate this, the POs on level two *traction battery*, *power train* and *car body* are independent from each other in the sense that it is possible to store each of these in a database, build a product system and perform an environmental assessment (e.g. LCA of a traction battery) without the necessity of considering the other two POs

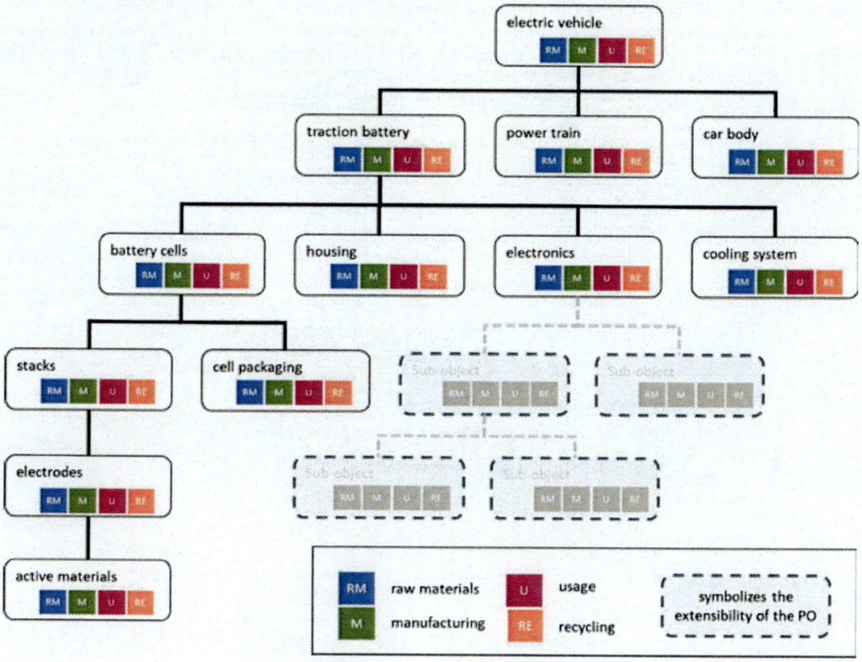

Fig. 4.21 Hierarchy of product objects in ICLCE. Each of the PO has the same structure as in Figs. 4.19 and 4.20

on the same level. Through joining a lower-level PO it is possible to extend the foreground system. The figure indicates this capability offered by the data structure with regard to the extendibility of the system boundary between the foreground and the background system of a PO. For instance, as shown in Fig. 4.21, the foreground includes the PO *electronics,* even though its components belong to the background system.

The attributes are, in general, environmentally related information such as material and energy flows resulting from the life cycle stages relationship models and performance properties that have an influence on other POs downstream.

Product system linking and evaluation

Notice in Fig. 4.22 that the POs at the different hierarchical levels are structured independent from each other but contain a link to the POs placed on a level below. The figure shows this schematically for the levels *electric vehicles*, *battery cells* and stacks (*ev, bc* and *stk* in the figure) for illustrative purposes. The figure shows furthermore the workflow to construct a PS after the POs has been created.

Once a query is submitted (e.g. EV1), the system starts the linking process to create a virtual PS. The PO EV1 is uniquely identified with a code (e.g. EV1_evuid), which enables it to retrieve all the lower level POs linked to it. First, the technosphere

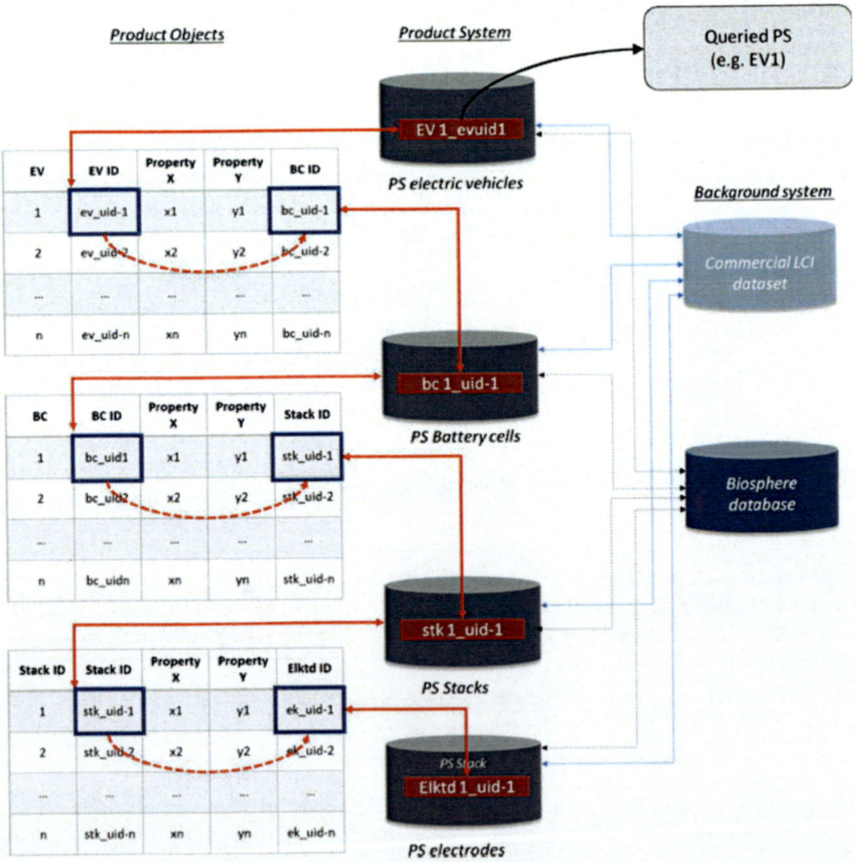

Fig. 4.22 Information flow between POs and PSs

and biosphere exchanges in the POs in lower levels are linked through the defined codes, searched for and extracted from the database. The objective here is to consider the defined spatial contexts while: (i) matching every technosphere exchange to the reference flow of a dataset in a commercial database that better represents it, and (ii) matching each biosphere exchange to the elements in the biosphere database. This is done automatically through the definition of mapping and relational query processes. Once the linking process is finished, the information is integrated into a specific data structure (e.g. a dictionary) containing the links to the background system, and the biosphere database is created and saved to a database. Datasets of product systems can now be searched for in the PS database using relational queries based on technical parameters (e.g. *all EV heavier than 1300 kg with a particular battery system,* etc.), and assessed using LCA routines as described by Heijungs and colleagues (Heijungs and Suh 2002; Suh and Huppes 2005; Mutel 2017).

4.4 Prototypical Implementation of an ICLCE for Traction Batteries

The presented concept has been prototypically implemented to illustrate the usability and advantages of the approach. The general technical implementation concept of the ICLCE concept for traction batteries is shown in Fig. 4.23. The figure gives an overview of the main information flow channels and the way in which the different system layers interact with each other.

As shown in the figure, through an interactive development environment, the system enables an interaction between the user and all the logic layers and main information flows: (a) definition of the foreground system, (b) definition of the background system, (c) definition of the spatial context, (d) product system generator,

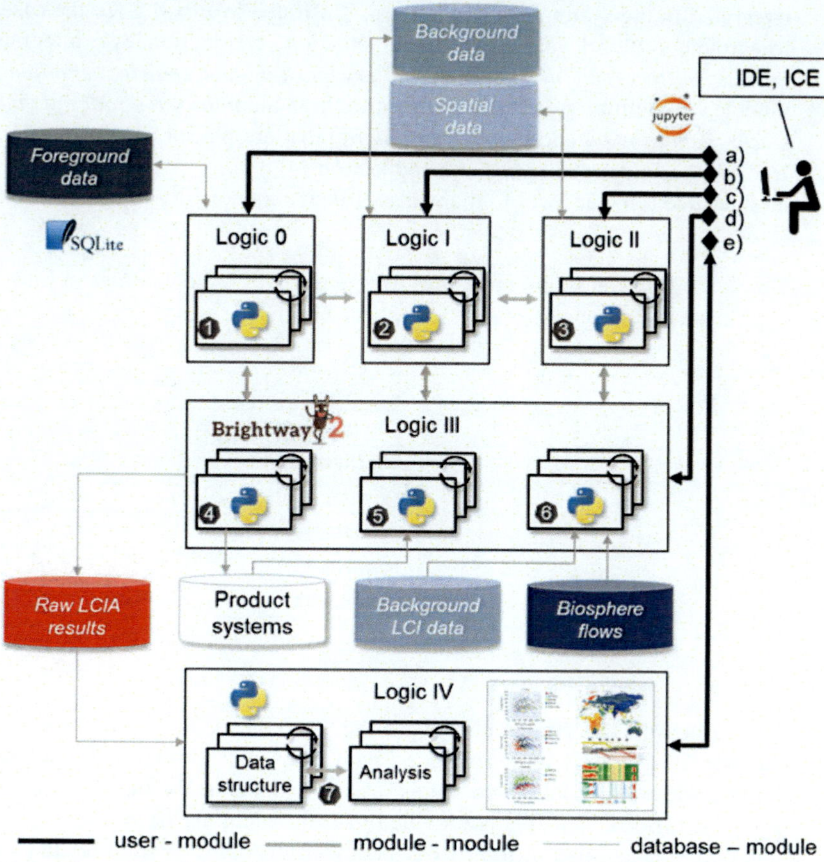

Fig. 4.23 Overview of the practical implementation of the concept

(e) definition of analytical goals. The figure also presents the different interactions between the logic layers and the datasets.

The system has been completely developed using python 3.6 (Van Rossum and Drake 2010) and diverse python packages for data wrangling, analysis and visualization such as pandas, numpy, scipy, matplotlib and seaborn. All logic layers were implemented as separated modules containing code scripts that integrate data and related functionality among them. This means that each layer module includes the required codes to send and receive information to or from other module layers and databases, but also to control information flows among the scripts in a module.

These modules are organized following a hierarchical folder structure that enables calling or importing functions from modules and scripts located at a level below the main level. This folder structure for the main level of modules is represented for illustrative purposes in Box 4.1. It contains one folder for each logic layer and each of the folders includes one script for each model that has been developed and integrated in the system. This is, in turn, illustrated in Box 4.2 for the cases of the foreground system layer. In this implementation version, this layer integrates three models: *battery_cell_definition.py, battery_pack.py and vehicle_definition.py.* The module cell definition is further related to three modules (*electrodes_definition.py, stacks_definition.py* and *cells_definition.py*) that are placed in a lower hierarchical level of the folder structure. The application of *__init__.py* file enables importing modules from lower level to make use of their functionality in models on a higher level.

Box 4.1 First level folder structure. Arrangement of logic modules in folders. Only for illustrative purposes

.../iclce/	
	foreground/...
	background/...
	spatial_context/...
	product_system/...
	analysis/...
	main.ipnb
	...

The possible user interaction with the developed ICLCE platform for traction batteries is abstractly represented in Fig. 4.24. A user interface fills the purpose of writing and reading data in the databases and to integrate the functions in the scripts. As shown in Box 4.1, this is provided to the user through a web-based interactive computing environment. In this case, Jupyter notebooks is used for the generation of code that can be instantly executed. (Kluyver et al. 2016).

The user has access to each of the layers containing integration. The models are accessible through commands in the Jupyter file at the highest folder level. These can be accessed with the purpose of calling the particular functions inside the scripts

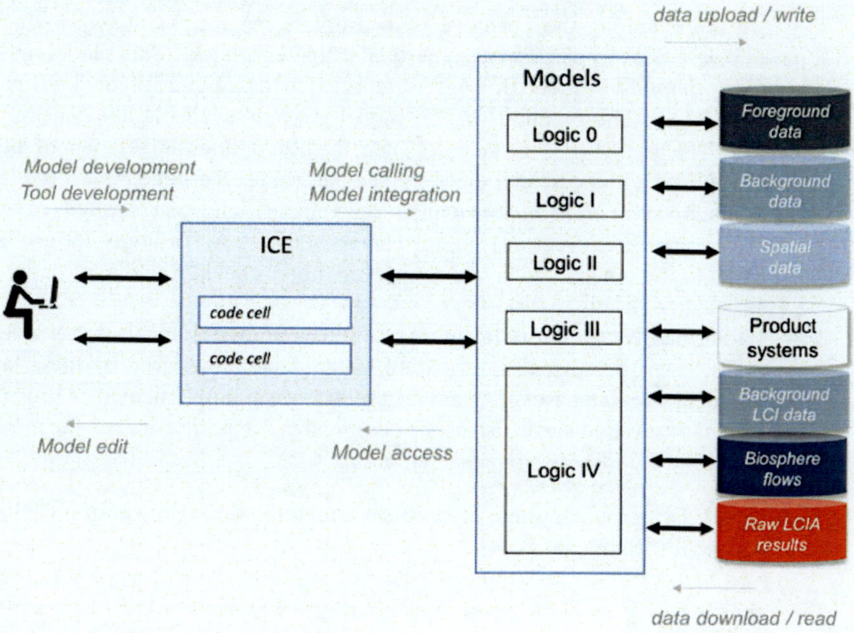

Fig. 4.24 User interface and interaction

that will form the foreground system (see a in Fig. 4.23), access and edit them or to store newly developed models within the layers. This is shown in the code snippet in Fig. 4.25 for a fictional example of a user interaction with the foreground system models through an ICE interface.

```
import foreground.vehicle_definition as ev
# calls logic layer foreground system

segment= "A"
# defines vehicle segment

battery_code=po_battery_packs["code"]
# defines vehicle segment

ev.calcuate_mass(battery_code, segment)
# calls the function to calculate the mass of the vehicle
# and transfers as input parameter the uuid code of a battery
# system and a defined segment
```

Fig. 4.25 Illustrative code snippet. Calling models and functions from lower folder levels while passing objects and parameters

In the example, the objective is to estimate the weight of the product object *EV* given the properties of a particular battery pack. The model in the file *vehicle_definition.py* (see Box 4.2) is imported to the jupyter notebook file *main.ipnb (code line 3)*. The functions in this model are now available and accessible to be edited. The user defines the vehicle. Then, the user defines the segment of the vehicle (*code line 6*) and the uuid code of a given battery pack product object (*code line 9*). The function *calculate_mass()* in the script vehicle_definition is called (*code line 12)* and the parameters previously defined are passed to it. Accordingly, the user is able to interact to the other logic layers containing models background and spatial_context (see b) and c) in Fig. 4.23).

Additionally, each logic layer provides programmed channels that enable placing queries to write and read information in the databases served by these layers. To implement the databases, SQLite (Hipp 2015) was implemented. All of the logic layers in the system serve also as the communication point between the modeler and the specific databases as represented in Fig. 4.23.

Box 4.2 Foreground system layer folder structure. Arrangement of model's scripts. Only for illustrative purposes

.../foreground/		
	__init__.py	
	battery_cells/...	
		__init__.py
		electrodes_definition.py
		stacks_definition.py
		cells_definition.py **manufacturing_processes/...** **cell_manufacturing.py**
		...
	battery_cell_definition.py **battery_pack.py**	
	vehicle_definition.py	

While the data is structured using standard formats, the application of regular queries enables reading and manually writing or uploading datasets. This is useful as it enables verification and reporting. This also speeds up consultation, as the user can directly access an already defined product system by browsing the database for specific queries. The objective of the user-module interaction in (see d) in Fig. 4.23) is to programmatically create product systems. These are based on the intermediate results and information created from the three previous logics, which are stored in the intermediate databases. The raw material metallurgical value chain (from mineral to metal) in the current implementation is considered part of the background system.

The inactive materials are also considered outside of the focus, and therefore part of the background system.

The user then proceeds to, also programmatically, build LCIs out of these product systems by linking the POs to the background system. In this implementation, the background system database implemented is ecoinvent 3.4. Also in this layer, the interaction allows the user to estimate their environmental impact given a particular methodology and impact category. The estimation of the life cycle environmental impacts has been implemented using the open source python framework for LCA called Brightway2 (Mutel 2017). As already mentioned, analytical models can also be integrated, edited or developed by the modeler (see e in Fig. 4.23).

The interaction between user and logic IV specifically enables this as seen in Fig. 4.24. Programmed scripts to wrangle, visualize and analyze large amounts of data are stored in this logic. For instance, a code to instantly generate LCA world heat-maps as in the work done by Egede (2016), product LCA heat-maps as in Cerdas and colleagues (2017) or preparing data to be used by innovative visualization devices and methodologies as described in (Juraschek et al. 2018; Kaluza et al. 2018, 2019). In the same way, analytical algorithms, for instance for the estimation of critical parameters through sensitivity analysis methodologies, can be easily integrated in this layer.

References

Allison J, Backman D, Christodoulou L (2006) Integrated computational materials engineering: a new paradigm for the global materials profession. JOM 58:25–27. https://doi.org/10.1007/s11837-006-0223-5

Allison J, Li M, Wolverton C, Su XM (2006) Virtual aluminum castings: an industrial application of ICME. Jom 58:28–35. https://doi.org/10.1007/s11837-006-0224-4

Butler H, Daly M, Doyle A et al (2016) The GeoJson format. Internet Eng Task Force 1–28

Cerdas F, Kaluza A, Erkisi-Arici S et al (2017) Improved visualization in LCA through the application of cluster heat maps. Procedia CIRP 00:732–737. https://doi.org/10.1016/j.procir.2016.11.160

Cerdas F, Thiede S, Herrmann C (2018) Integrated computational life cycle engineering—application to the case of electric vehicles. CIRP Ann 67:25–28. https://doi.org/10.1016/j.cirp.2018.04.052

Cohen M (1976) Unknowables in the essence of materials science and engineering. Mater Sci Eng 25:3–4

Egede P (2016) 'Environmental assessment of lightweight electric vehicles 94–98

Groen EA, Bokkers EAM, Heijungs R, de Boer IJM (2017) Methods for global sensitivity analysis in life cycle assessment. Int J Life Cycle Assess 22:1125–1137. https://doi.org/10.1007/s11367-016-1217-3

Hauschild MZ, Herrmann C, Kara S (2017) An integrated framework for life cycle engineering. Procedia CIRP 61:2–9. https://doi.org/10.1016/j.procir.2016.11.257

Heijungs R, Suh S (2002) The computational structure of life cycle assessment. Springer, Netherlands, Dordrecht

Hipp R (2015) SQLite (Version 3.8.10.2)

Horstemeyer MF (2012) Integrated computational materials engineering (ICME) for metals—using multiscale modeling to invigorate engineering design with science. Wiley Inc., Hoboken, NJ, USA

Horstemeyer MF, Sahay S (2018) Definition of ICME. In: Horstemeyer MF (ed) Integrated computational materials engineering (ICME) for metals. Wiley Inc., Hoboken, NJ, USA, pp 1–17

Juraschek M, Büth L, Cerdas F et al (2018) Exploring the potentials of mixed reality for life cycle engineering. Procedia CIRP

Kaluza A, Juraschek M, Büth L et al (2019) Implementing mixed reality in automotive life cycle engineering. Procedia CIRP 80:717–722. https://doi.org/10.1016/j.procir.2019.01.078

Kaluza A, Gellrich S, Cerdas F et al (2018) Life cycle engineering based on visual analytics. Procedia CIRP

Keim D, Andrienko G, Fekete J-D et al (2005) Visual analytics: definition, process, and challenges. Information visualization. Springer, Berlin, pp 154–175

Kluyver T, Ragan-Kelley B, Pérez F et al (2016) Jupyter notebooks—a publishing format for reproducible computational workflows. Position Power Acad Publ Play Agents Agendas 87–90. https://doi.org/10.3233/978-1-61499-649-1-87

Leffingwell D, Widrig D (2000) Managing software requirements: a unified approach. Addison-Wesley Longman Publishing Co., Boston, MA USA

Liu S (2008) Global sensitivity analysis: the primer by Andrea Saltelli, Marco Ratto, Terry Andres, Francesca Campolongo, Jessica Cariboni, Debora Gatelli, Michaela Saisana, Stefano Tarantola

Mutel C (2017) Brightway: an open source framework for life cycle assessment 47:11–12. https://doi.org/10.21105/joss.00236

NAE (2008) Integrated computational materials engineering. National Academies Press, Washington, D.C.

Nattermann R, Anderl R (2013) The W-model using systems engineering for adaptronics. Procedia Comput Sci 16:937–946. https://doi.org/10.1016/j.procs.2013.01.098

Olson GB (1997) Computational design of hierarchically structured materials. Science (80-)277:1237–1242. https://doi.org/10.1126/science.277.5330.1237

Panchal JH, Kalidindi SR, McDowell DL (2013) Key computational modeling issues in integrated computational materials engineering. CAD Comput Aided Des 45:4–25. https://doi.org/10.1016/j.cad.2012.06.006

Saltelli A, Annoni P, Azzini I et al (2010) Variance based sensitivity analysis of model output. Design and estimator for the total sensitivity index. Comput Phys Commun 181:259–270. https://doi.org/10.1016/j.cpc.2009.09.018

Schmitz GJ, Prahl U (2016) Handbook of software solutions for ICME

Schmitz GJ, Prahl U (2012) Introduction. Integrative computational materials engineering. Wiley-VCH Verlag GmbH & Co. KGaA, Weinheim, Germany, pp 1–20

Steubing B, Wernet G, Reinhard J et al (2016) The ecoinvent database version 3 (part II): analyzing LCA results and comparison to version 2. Int J Life Cycle Asses

Suh S, Huppes G (2005) Methods for life cycle inventory of a product. J Clean Prod 13:687–697. https://doi.org/10.1016/j.jclepro.2003.04.001

Taylor CD, Lu P, Saal J et al (2018) Integrated computational materials engineering of corrosion resistant alloys. npj Mater Degrad 2:6. https://doi.org/10.1038/s41529-018-0027-4

TMS (2013) Integrated computational materials engineering (ICME): implementing ICME in the aerospace, automotive and maritime industries. Miner Met Mater Soc. https://doi.org/10.1109/ICALT.2007.59

Van RG, Drake FL (2010) Python tutorial. History 42:1–122. https://doi.org/10.1111/j.1094-348X.2008.00203_7.x

Vandepaer L, Cloutier J, Bauer C, Amor B (2018) Integrating batteries in the future Swiss electricity supply system: a consequential environmental assessment. J Ind Ecol. https://doi.org/10.1111/jiec.12774

Velten K (2008) Mathematical modeling and simulation. Wiley-VCH Verlag GmbH & Co, KGaA, Weinheim, Germany

Wernet G, Bauer C, Steubing B, et al (2016) The ecoinvent database version 3 (part I): overview 991 and methodology. Int J Life Cycle Assess

Chapter 5
Exemplary Application: Analysis of Variability in the LCE of Batteries for Electric Vehicles

Contents

The ICLCE concept serves as a methodology to develop life cycle engineering models through the coupling of discipline specific models. The implemented software-prototype a platform to integrate discipline specific models representing specific unit processes in the life cycle of a traction battery. This chapter presents the complete application process, starting with setting of the research question and the definition of the system boundaries, through describing the models and models coupling strategy and presenting and analysis of the results. The general theme of the case study is the analysis of the influence of technological and spatial variability on the LCA of lithium ion batteries and their application in electric vehicles. In this regard, the objective is to demonstrate the applicability of the concept to support LCE in the field of electromobility. In the first part of the chapter, the effect of relevant technical parameters with regard to the geometry and materials of a battery cell, its production processes and spatial context are investigated. The second part focuses on the usage stage of an EV. Specifically, parameters regarding the mass of vehicle, as well as variable geographical and temporal factors affecting the energy consumption, are explored in context.

F. Cerdas, *Integrated Computational Life Cycle Engineering for Traction Batteries*, Sustainable Production, Life Cycle Engineering and Management,
https://doi.org/10.1007/978-3-030-82934-6_5

5.1 Introduction

Compared to ICEV, the potential environmental impact caused by EV is much more sensitive to the variability of the technical parameters defining their performance and to the external factors providing the context in which they operate as represented in Fig. 5.1.

Two reasons can be mentioned in this regard. First, the environmental impact of ICEV is predominantly located in the usage stage, driven by the amount of fossil fuels consumed and the tailpipe emissions generated. This is, in contrast to an EV, to a very large extent independent from the place or region where the vehicle is being operated. On the one hand, energy conversion efficiency in ICEV is very low compared to EVs, resting significance to the leverage of other factors during the operation of the vehicle. On the other hand, the energy required for heating is exchanged by the engine to the cabin. If the electricity mix provided to charge a traction battery includes a high share of fossil-based energy sources, then GHG-based environmental emissions such as GWP or fossil depletion potential (FDP) will be in the same way predominantly located in the usage stage of the vehicle. The difference is that the energy conversion efficiency of the EV is much higher, which in turns rises the leverage of the external parameters (e.g. driving profile and external temperatures) influencing the amount of energy consumed per km. These parameters are additionally variable in time. In contrast to the WTW process chain of an ICE, the composition of electricity usually changes dynamically, including periods containing low and high shares of renewable energies which has further repercussions on the usage-stage related impact.

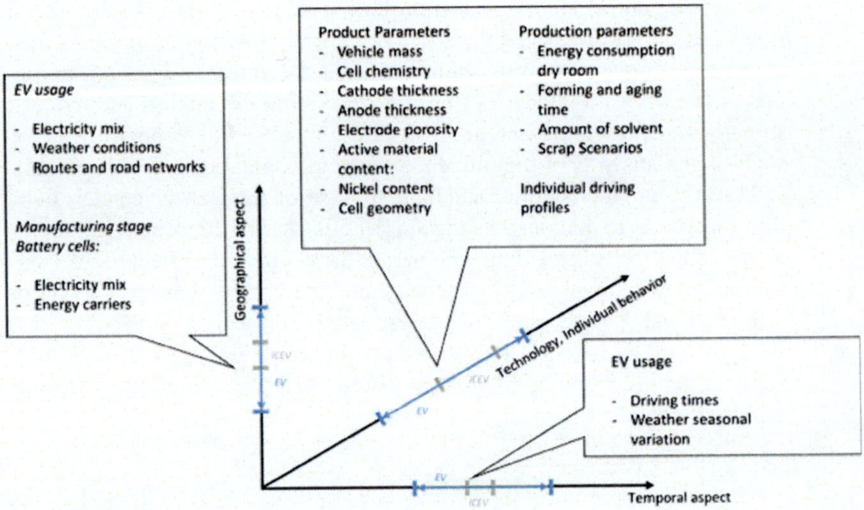

Fig. 5.1 General sources of variability in LCIA results of EVs considered in the current implementation of the ICLCE approach

Second, if the carbon intensity of the electricity mix in the usage stage of an EV is comparably low, the life cycle carbon-driven environmental impact is distributed throughout the other life cycle stages in which the battery system becomes a much more significant role. Specifically, its cradle to gate production emerges as the largest hotspot of impacts, driven by the production of the required materials and the manufacturing phase of the battery cells. In this regard, the diversification of technical parameters and scenarios, as well as the different geographical and temporal contexts, influence the final environmental footprint.

The intention of the case study presented in this chapter is to provide an example of the usability of the modelling approach as described in Chap. 4 for the quantification of LCA results of traction batteries in EV applications, considering all these aforementioned sources of variability. Specifically, the aim of the case study is to research the spread of the range in LCIA results of a EV traction battery, while considering a high degree of variability in the most relevant technological parameters and spatial circumstances associated to its life cycle.

For this purpose, the study is divided into two parts. First, a cradle to gate integrated model will be developed for the assessment of traction batteries. Second, a model for the analysis of the use stage will be fed with the traction batteries created in the first step.

An overview of the models to be integrated is shown in Fig. 5.2. Three sets of models were adjusted and integrated in the foreground system layer. First, the definition of the batteries geometrical and physical properties is carried-out by coupling

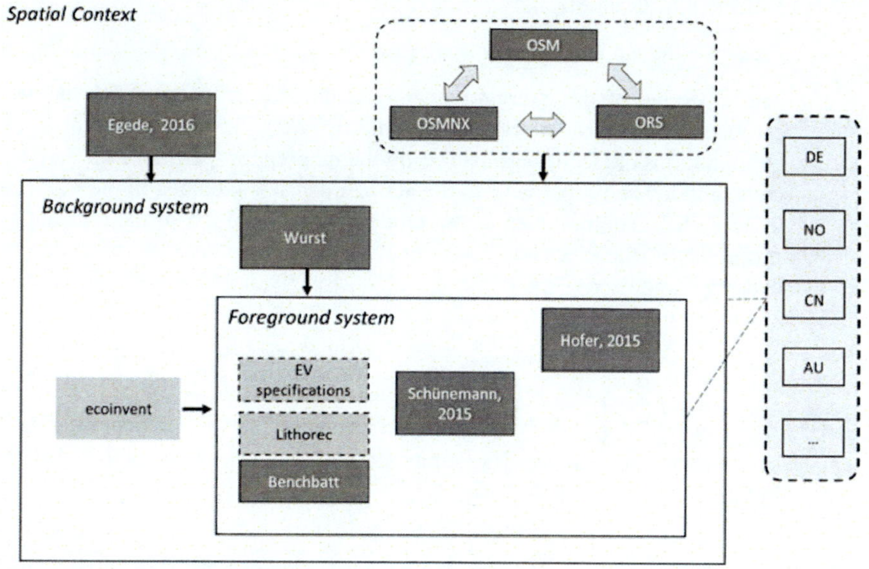

Fig. 5.2 System boundaries definition in the prototypic implementation of the ICLCE for traction batteries

an energy-mass model from material to cell with geometrical constraints from cell forms, battery systems and vehicle specifications.

The models integrated in the current implementation of the system intend to supporting a study that analyzes the influence of:

i. product parameters at cell level such as electrode thickness and cell chemistry
ii. production parameters such as energy consumption in the manufacturing stage
iii. vehicles size and mass
iv. the electricity mix during the manufacturing stage
v. the electricity mix during the usage stage
vi. ambient temperature during the usage stage.

A model coupling matrix defined for the following implementation is included in the appendix.

5.1.1 Implemented Models in the Foreground System Layer

The energy-mass model developed within the project Benchbatt (Schmuch et al. 2018; Cerdas et al. 2018b; Betz et al. 2019) was further developed and complemented with linear models. These linear models used the data gathered during disassembly experiments by the LithoRec project (Cerdas et al. 2018a), and other geometric data regarding available space for traction batteries in commercial electric vehicles, collected in (Grunditz and Thiringer 2016; Schmuch et al. 2018).

Energy-mass model from electrode to stack

The objective of the energy-mass model developed in Benchbatt is the physical estimation of the volumetric and gravimetric energy at stack level of a set of feasible combinations of electrodes, using only the inherent properties of the active materials considered. The properties and the active materials included in the model are listed in Table 5.1. The first step of the model is the definition of the relevant geometric properties at the electrode level. In this regard, the estimation of the volumetric capacity of the active material $c_{vol|am}\left[\frac{\text{mAh}}{\text{cm}^3}\right]$ is given in the model by the equation:

$$c_{vol|am} = q_{am} \cdot \rho_{am|crystal} \tag{5.1}$$

where $q_{am}\left[\frac{\text{mAh}}{\text{g}}\right]$ is the specific capacity of the material and $\rho_{am|crystal}\left[\frac{\text{g}}{\text{cm}^3}\right]$ its crystallographic density. The composite electrode density $\rho_{com}\left[\frac{\text{g}}{\text{cm}^3}\right]$ is then given by:

$$\rho_{com} = \frac{1}{\frac{mf_{am}}{\rho_{am|crystal}} + \frac{mf_{bin}}{\rho_{bin}} + \frac{mf_{ca}}{\rho_{ca}}} \tag{5.2}$$

Table 5.1 Exemplary active materials included in Benchbatt energy-mass model (Schmuch et al. 2018; Cerdas et al. 2018b; Betz et al. 2019)

Active material	Abbreviation	Practical discharge capacity (mAh/g)	Average discharge potential versus Li/Li + (V)	Crystallographic density (g/cm^3)
$LiNi_{0.33}Mn_{0.33}Co_{0.33}O_2$	111-NMC	160	3.7	4.7
$LiNi_{0.4}Mn_{0.4}Co_{0.2}O_2$	442-NMC	165	3.7	4.7
$LiNi_{0.5}Mn_{0.3}Co_{0.2}O_2$	532-NMC	170	3.7	4.7
$LiNi_{0.6}Mn_{0.2}Co_{0.2}O_2$	622-NMC	180	3.7	4.7
$LiNi_{0.8}Mn_{0.1}Co_{0.1}O_2$	811-NMC	200	3.7	4.7
LiC_6	Gr	360	0.1	2.2
$LiFePO_4$	LFP	160	3.3	3.58
Li	Li_0	3884	0	0.53
Li	Li/2	1942	0	0.53
Li_2O_2	Li_2O_2	1165	2.6	2.31
Li_2S	Li_2S	1650	2.1	1.66
$LiCoO_2$	LCO	150	3.8	5
$LiNi_{1.5}Mn_{0.5}O_4$	LNMO	140	4.7	4.4
$Li_{1.2}Ni_{0.16}Mn_{0.56}Co_{0.08}O_2$	LR-NMC	250	3.35	4.25
$Li_4Ti_5O_{12}$	LTO	160	1.55	3.43
$LiNi_{0.8}Al_{0.15}Co_{0.05}O_2$	NCA	200	3.7	4.7
$LiSi_x$	Si-1000	1000	0.4	2.2

where mf_{am}, mf_{bin}, mf_{ca} [dimensionless] are the mass fractions of active material, binder and conductive additive respectively and $\rho_{bin}\left[\frac{\text{g}}{\text{cm}^3}\right]$ and $\rho_{ca}\left[\frac{\text{g}}{\text{cm}^3}\right]$ the densities of the binder and the conductive additive respectively. The density of the active material fraction in the electrode $\rho_{el|am}\left[\frac{\text{g}}{\text{cm}^3}\right]$ considering the composite's porosity is then estimated as:

$$\rho_{el|am} = mf_{am} \cdot \rho_{com}(1 - \varepsilon) \tag{5.3}$$

where ε [dimensionless] is the porosity of the electrode. The electrode density including the electrolyte $\rho_{el}\left[\frac{\text{g}}{\text{cm}^3}\right]$ can now be estimated as:

$$\rho_{el} = (1 - \varepsilon) \cdot \rho_{com} + \varepsilon \cdot \rho_{elek} \tag{5.4}$$

where $\rho_{elek}\left[\frac{\text{g}}{\text{cm}^3}\right]$ is the density of the electrolyte.

The electrode's volumetric capacity $c_{vol|el}\left[\frac{\text{mAh}}{\text{cm}^3}\right]$ is given by:

$$c_{vol|el} = q_{am} \cdot \rho_{el|am} \tag{5.5}$$

Finally, the specific material gravimetric $E_{mat|grav}\left[\frac{\text{Wh}}{\text{kg}}\right]$ and volumetric energy $E_{mat|vol}\left[\frac{\text{Wh}}{\text{l}}\right]$ are calculated as:

$$E_{mat|grav} = q_{am} \cdot V_{am|vsLi/Li} \tag{5.6}$$

$$E_{mat|vol} = c_{vol|el} \cdot V_{am|vsLi/Li} \tag{5.7}$$

where $V_{am|vsLi/Li}$ [V] is the average discharge potential.

The second step of the model is the definition of the volumetric and gravimetric energy and other relevant geometrical parameters at the stack level. To start, the area mass of the positive electrode, $Am_{pos}\left[\frac{\text{mg}}{\text{cm}^2}\right]$, is calculated as:

$$Am_{pos} = t_{pos} \cdot \rho_{pos|am} \tag{5.8}$$

where t_{pos} [μm] is the thickness of the active material coating and $\rho_{pos|am}$ is given by the Eq. (5.3). Further, the positive and negative area capacities $c_{area|pos}\left[\frac{\text{mAh}}{\text{cm}^2}\right]$ and $c_{area|neg}\left[\frac{\text{mAh}}{\text{cm}^2}\right]$ are obtained from:

$$c_{area|pos} = t_{pos} \cdot c_{vol|pos} \tag{5.9}$$

$$c_{area|neg} = c_{area|pos} \cdot np_{ratio} \tag{5.10}$$

where $c_{vol|pos}$ is given by the Eq. (5.5) and np_{ratio} [dimensionless] is the capacity ratio between the positive and negative electrode. With this, the thickness of the negative electrode t_{neg} [μm] is calculated as:

$$t_{neg} = \frac{c_{area|neg}}{c_{vol|neg}} \tag{5.11}$$

and with this the area mass $Am_{neg}\left[\frac{\text{mg}}{\text{cm}^2}\right]$ of the negative electrode is then estimated as:

$$Am_{neg} = t_{neg} \cdot \rho_{neg|am} \tag{5.12}$$

with $\rho_{neg|am}$ being calculated by the Eq. (5.3). The total thickness of the stack t_{stack} [μm] can now be estimated as presented in Fig. 5.3. In this regard, the total stack thickness is given by:

$$t_{stack} = t_{pos} + t_{neg} + \frac{t_{pos|coll}}{2} + \frac{t_{neg|coll}}{2} + t_{sep} \tag{5.13}$$

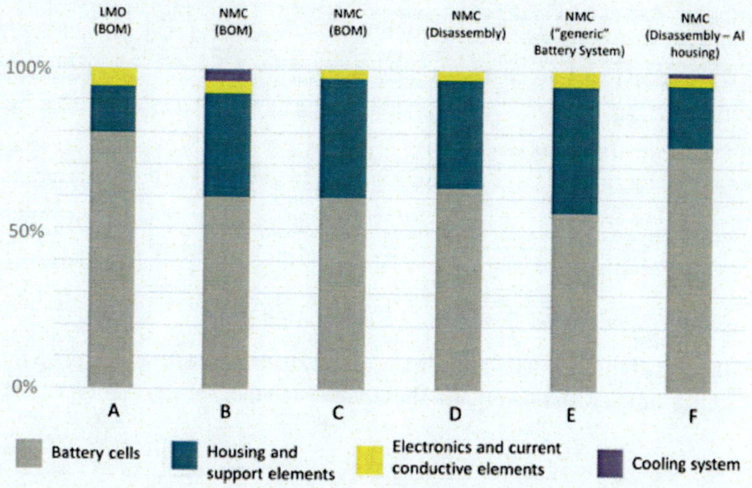

Fig. 5.3 Share of active against inactive materials contained in a traction battery system throughout available data from publications and disassembly experiments. **a** LMO (Notter et al. 2010), **b** NMC (Ellingsen et al. 2014), **c** NMC (Hawkins et al. 2013), **d–f** (Kwade et al. 2016; Cerdas et al. 2018a)

being $t_{pos|coll}$ [μm], $t_{neg|coll}$ [μm] and t_{sep} [μm] the thickness of the electrodes' current collectors and the separator respectively. Given the values in Table 5.2, it is then then possible to calculate the stack's areal mass Am_{stack} $\left[\frac{\text{mg}}{\text{cm}^2}\right]$.

The nominal cell voltage V_{nom} [V] is calculated as the difference between the average discharge potential of both active materials:

$$V_{nom} = V_{pos|vs\,Li/Li} - V_{neg|vs\,Li/Li} \tag{5.14}$$

Table 5.2 Exemplary inactive materials included in IC-LCE

Cell inactive material	Thickness (cm)	Specific mass (g/cm^2)	Density (g/cm^3)	Porosity
Pouch foil	0.012	0.018	2.000	
Binder			1.800	
Positive current collector (Al; per coated side)	0.002	0.003	2.700	
Negative current collector (Cu; per coated side)	0.001	0.004	8.960	
Liquid electrolyte			1.200	
Separator	0.002	0.001	0.950	0.400
Conductive additive			2.250	

Finally, the total stack volumetric $E_{stack|vol}$ $\left[\frac{\text{Wh}}{\text{l}}\right]$ and gravimetric energy $E_{stack|grav}$ $\left[\frac{\text{Wh}}{\text{kg}}\right]$ are given:

$$E_{stack|vol} = V_{nom} \cdot \frac{c_{area|pos}}{t_{stack}} \tag{5.15}$$

$$E_{stack|grav} = V_{nom} \cdot \frac{c_{area|pos}}{Am_{stack}} \tag{5.16}$$

Energy-mass model from cell to battery pack

In this part of the model, the properties at stack level are up-scaled to match a specific cell geometry. In this regard, the number of stacks n_{stack} [units] is estimated as:

$$n_{stack} = \frac{t_{cell}}{t_{stack}} \tag{5.17}$$

In this way, it is possible to estimate mass and energy properties of commercial cell geometries containing the stacks defined in the previous step. The implemented version of the model includes a list of specifications for some commercial battery cells in the database. A summary of geometrical constraints of battery cells is given in Table 5.3. To further upscale the mass and energy properties of cells to the battery system level, two design constraints are set in this part of the model. First, the available volume for active material in a commercial vehicle and the share of active to inactive material in a battery pack. Regarding the first constraint, the model in this section is linked to a database of commercial electric vehicles providing information about the nominal capacity, mass and other properties of a current commercial traction battery system. Regarding the second constraint, an analysis was done using several published material inventories for traction batteries, as well as data resulting from the disassembly processes of spent traction batteries from previous projects (Cerdas et al. 2018a). The objective of the analysis was to obtain a rough estimation of the share between the active (battery cells) and the inactive (housing, cooling components and electronics) materials of a battery system.

Table 5.3 Exemplary geometries of commercial battery cells integrated in the ICLCE

Commercial Cell	Volume (l)	Mass (kg)	Length (mm)	Height (mm)	Width (mm)
Li Tec	0.60	1.25	208	248	11
Samsung SDI-60	0.97	1.80	173	125	45
Toshiba	0.85	1.70	171	113.5	43.8
Samsung SDI-94	0.97	2.00	173	125	45
LG Chem	0.49	0.86	290	216	7.82
Panasonic–Sanyo	0.33	0.72	147	90	25
AESC (NEC)	0.40	0.80	290	216	6.39

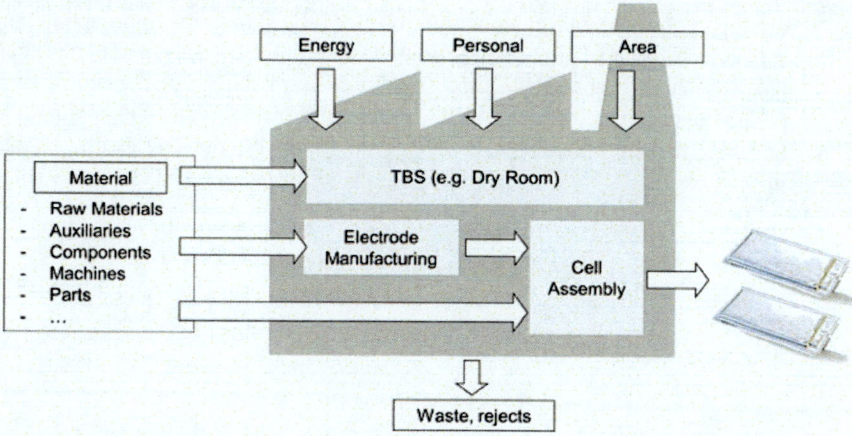

Fig. 5.4 Basic structure of the manufacturing cost model from Schünemann integrated ICLCE. Based on Schünemann (2015b)

A summary of the results is shown in the Fig. 5.3. Three inventories taken from the peer-reviewed literature relied on information given as a bill of materials (BOM), and the remaining inventories were retrieved from published data on recycling and from disassembly experiments performed in past research projects. A variation in the share of active material among the battery systems analyzed was identified (grey color). It ranges from around 55–80%, and essentially depends on the material selected for the structural housing of the battery system. A deductive modelling is then performed to calculate the amount of battery cells required in the battery pack.

Cell manufacturing model

The prototypic implementation of the system integrates parts of the model developed by Schünemann and documented in (Schünemann 2015b) for the evaluation of the manufacturing stage of battery cells. As represented in Fig. 5.4, the developed model follows a bottom-up modelling approach, in which specific information at the machine, process chain and factory levels are linked in a reverse material flow modelling approach. This departs from the quantification of the production throughput which is then transferred backwards to each of the manufacturing processes. The throughput flow is then coupled to the cells defined in the previous models, which then generates a material and energy flow analysis.

Usage phase energy consumption

The module implemented to estimate energy consumption is composed of a model that estimates the mechanical energy demand and a model for the quantification of auxiliary energy demand and standstill losses and charging losses.

For the estimation of the mechanical energy demand, the approach followed by Hofer was implemented (Hofer 2014a). Hofer applied the software Advisor (Gao et al. 2007) in order to estimate the energy consumption required to drive a given driving cycle. The energy consumption from the simulation approach is parameterized into A, B and C for the traction phase and A^1, B^1 and C^1 for the regeneration phase, so that it is possible to express it as a decoupled function depending on physical parameters such as the mass of the vehicle (m), frontal area (A_f), rolling coefficient (c_r) and drag coefficient (c_d). In this regard, for the traction energy (E_{trac}) this results in:

$$E_{trac} = \frac{1}{x_{tot}} \int\limits_{P_w>0} (P_a + P_r + P_k)dt \tag{5.18}$$

which can be further expressed as:

$$E_{trac} = \frac{1}{x_{tot}} \int\limits_{P_w>0} \left(\frac{1}{2} \cdot \rho \cdot c_d \cdot A_f \cdot v^3 + c_r \cdot m \cdot g \cdot v + m \cdot a \cdot v \right) dt \tag{5.19}$$

This is further factorized as:

$$E_{trac} = A \cdot c_d \cdot A_f + B \cdot c_r \cdot m + C \cdot m \tag{5.20}$$

with the coefficients A, B and C defined as:

$$A = \frac{\rho}{2 \cdot x_{tot}} \int\limits_{P_w>0} v^3 dt \tag{5.21}$$

$$B = \frac{g}{x_{tot}} \int\limits_{P_w>0} v dt \tag{5.22}$$

$$C = \frac{1}{x_{tot}} \int\limits_{P_w>0} a \cdot v dt \tag{5.23}$$

Similarly, the regeneration energy (E_{regen}) and dissipative energy (E_{diss}) are expressed as:

$$E_{regen} = A' \cdot c_d \cdot A_f + B' \cdot c_r \cdot m + C' \cdot m \tag{5.24}$$

$$E_{diss} = \left(A + A'\right) \cdot c_d \cdot A_f + \left(B + B'\right) \cdot c_r \cdot m \tag{5.25}$$

From the simulative approach followed with Advisor, the values obtained by Hofer for the WLTP driving cycle are summarized in Table 5.4.

In the case of the modelling of the energy demand by the auxiliaries' inputs from the eLCAr guidelines (Duce et al. 2013), the power demand of a vehicle's

Table 5.4 Parameterized factors for the estimation of energy consumption of vehicles for different segments of the WLTP driving cycle (Hofer 2014b)

	Unit	Average	Low	Medium	High	Very high
A		27,072	3410	10,079	23,314	48,992
B		774	586	716	819	839
C		12.6	19.7	18.3	10.5	8.4
A′		3692	2065	3068	2350	5826
B′		207	395	265	162	142
C′		−12.6	−19.7	−18.3	−10.5	−8.4

heater is 5 kW for temperatures below 10 °C and 2.5 kW for temperatures varying between 10 and 15 °C. The air conditioning, in turn, has a nominal power demand of 0.5 kW for temperatures between 20 and 25 °C and 1 kW for temperatures above 25 °C. Additionally, the model follows the eLCAr guidelines to estimate the energy demand by the different electric devices. In the model, factors can be adjusted so that it is possible the select different nominal power demand values and mean use ratios.

5.1.2 *Implemented Models in the Spatial Context Layer*

As represented in Fig. 5.5, GeoCoder (Carriere 2018), a python geocoding library in combination with the services sourced by Open Street Maps (OSM) (OpenStreetMap contributors 2015), was used to retrieve spatial information such as coordinates, bounding boxes and addresses. To retrieve information of routes between two or more points, including distance, average speed, type of road (city, highway) and elevation, the services provided by Graphhopper (Graphhopper contributors 2018) and Open Route Service ORS (GIScience 2008) were linked through application programming interfaces APIs. As reviewed in Chap. 2, the local temperature contributes largely to the environmental impact of an EV, as it influences its energy consumption per kilometer driven. The ICLCE integrates a module that enables linking hourly temperature profiles for all the months in the year in a particular place. This was done by implementing the methodology developed by Egede (2017). In her work, the Köppen-Geiger climate zone classification was used to develop monthly and hourly temperature profiles for each of the regions.

The Köppen-Geiger classification system for the period of 1986 through 2016 is shown in Fig. 5.6 (Beck et al. 2018). The system classifies all climatic regions into five main groups: A (tropical), B (dry), C (temperate), D (Continental) and E (polar). These main groups are further classified based on temperature and precipitation patterns resulting in a total of 30 climate zones (Kottek et al. 2006; Beck et al. 2018). As shown, these climate zones are defined as polygons limited by a set of coordinates points.

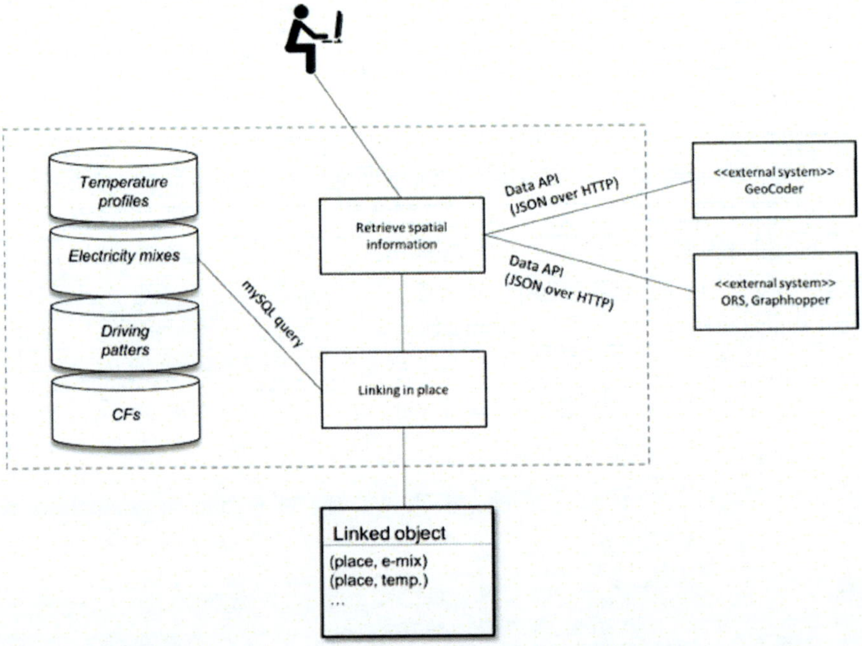

Fig. 5.5 Connection between internal and external system through API

In the IC-LCE methodology, this information is structured in a database. This makes it possible to retrieve the climate zone to which a specific point on Earth belongs to. Following the methodology proposed by Egede (2017), temperature profiles showing monthly and hourly variations were then created for each of the zones, and structured in the database. Accordingly, once a query is submitted for a specific city, the system matches the city to a climate zone and calculates the temperature profiles. In addition, a process linking the entire life cycle of an EV to the specific electricity mix being supplied at the point of consumption was implemented. This is done through linking a given place to its specifications in the database regarding a particular distribution of energy sources per unit of output.

Forecasts of electricity mixes are stored in the database and can be re-defined by the user for reasons such as generating future scenarios, for instance. Additionally, the current implementation of the system contains collections of current target electricity mixes based on data from the international energy agency, for example. The linking process is done following a set of rules as shown in Fig. 5.8. First, it is determined if the place is linked to a forecasted or targeted electricity mix, according to the scope of the analysis. This means, for example, that if the temporal boundaries defined for the study concerns a moment in the future, the system will link the process in the foreground system to a target or to a forecasted electricity mix for a particular region. If the study does not require linking the process to a future energy mix scenario, the activity will be linked in place to the respective current electricity mix. In this

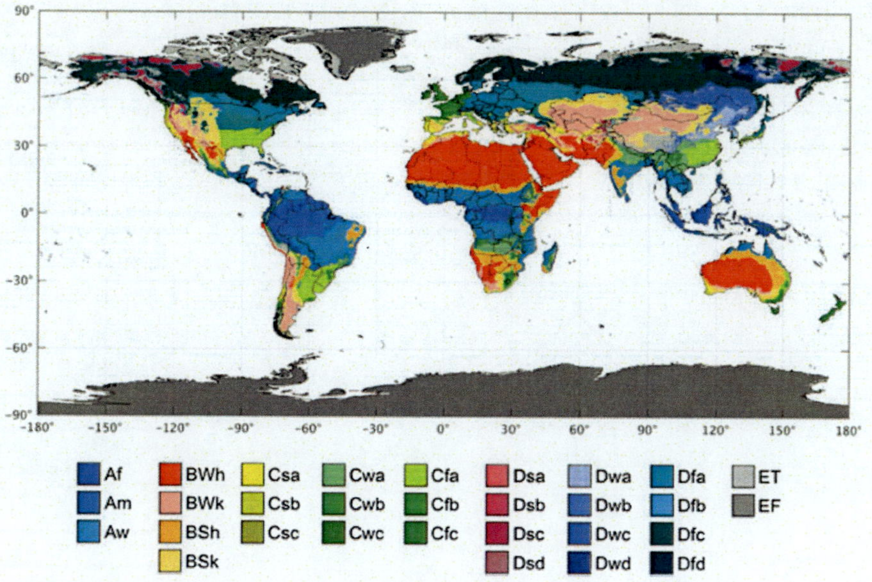

Fig. 5.6 Köppen-Geiger classification from 1986 to 2016 (Beck et al. 2018)[1]

[1] For a more detailed description of the Köppen-Geiger classification system please refer to the work done by Kottek and colleagues and Beck and colleagues (Kottek et al. 2006; Beck et al. 2018) and a description of methodology to produce build the temperature profiles in the work done by Egede (2017).

regard, the algorithm will first proof if there is a specification for the electricity mix for the particular place in the database. This distribution, along with its time resolution, is built for all the countries and regions for which there is data available. If this information is not available, the system will link the place automatically to an electricity mix in ecoinvent. The electricity mixes in the database are stored in three different time resolutions: daily average, monthly average and annual average. Thus, for some regions it is possible to link a process to monthly electricity averages or even to real time information through the application of APIs for services such as ©Tomorrow.[2] The added value of doing this is that it enables linking unit process that are sensitive to a temporal variability. For instance, the environmental impact of an EV depends strongly on the environmental impact embedded in the electricity being supplied to it during its use phase.

The composition of this electricity largely depends on climatic factors such as wind availability and speed, and solar availability and radiation, to name a few. These factors, together with demand cycles, make the mix of sources of electricity change

[2] https://www.electricitymap.org.

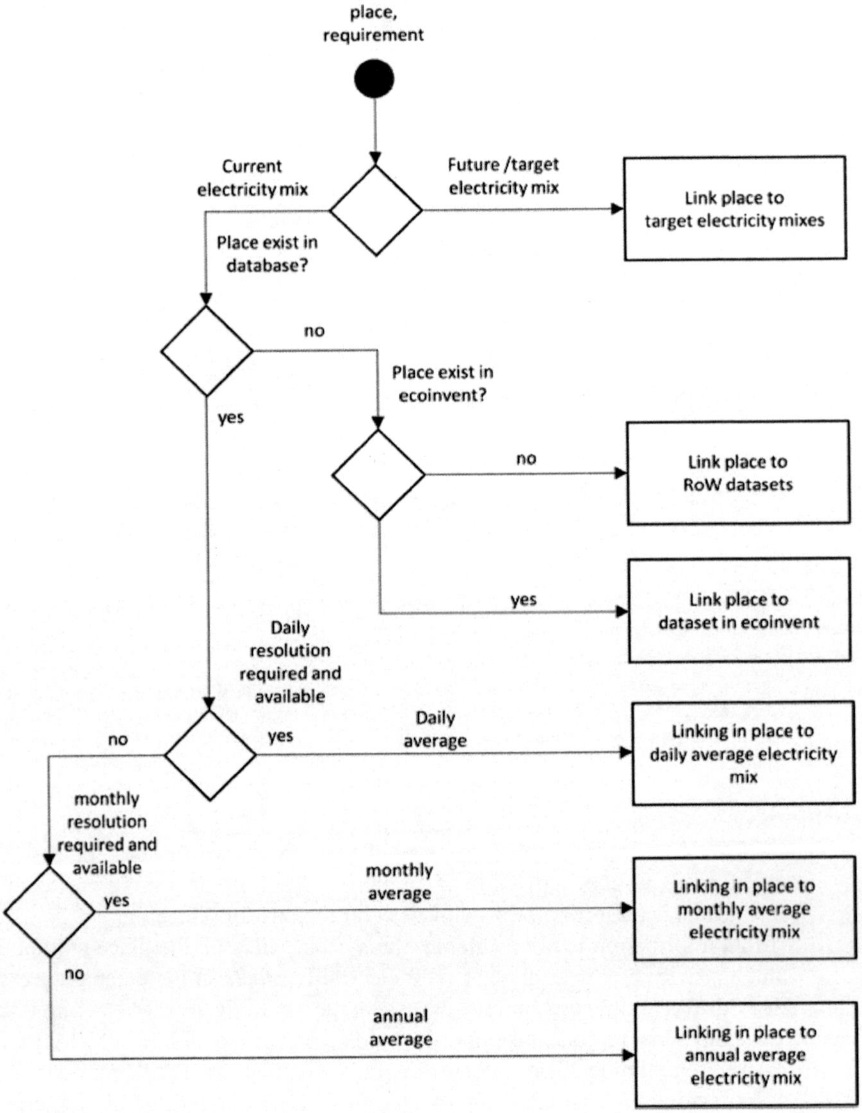

Fig. 5.7 Linking process for energy mixes and background datasets

daily and even hourly, resulting in the important task of capturing this variability in the modelling of the life cycle inventory.

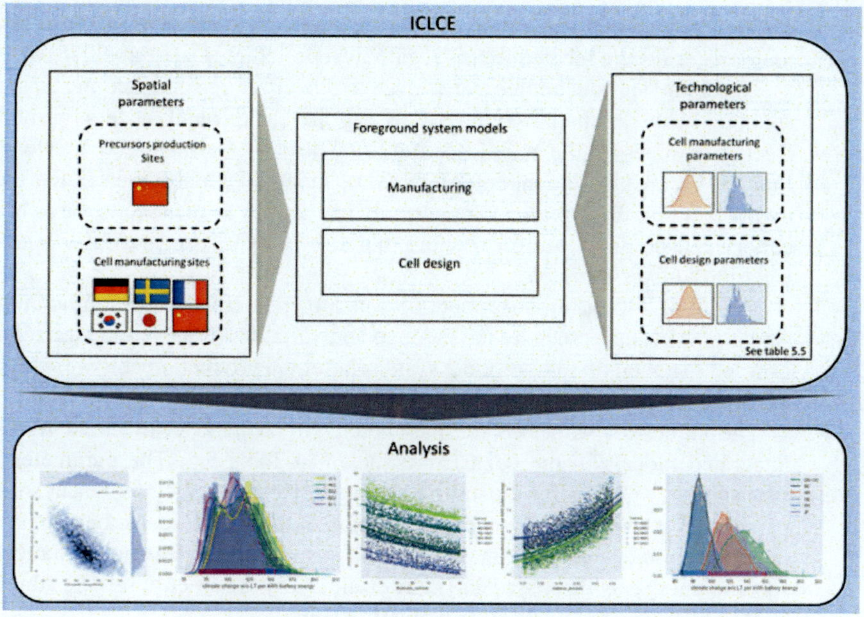

Fig. 5.8 Methodology followed for the case study #1

5.2 Case Study

The case study presented in the following sections is composed of two parts. First, an analysis that focuses on the complexity of the cradle to gate LCIA of traction batteries is presented. In the second part, the focus of the case study is the analysis of the usage stage of EVs. In both sections of the case study, the models integrated in the general LCE model are parameterized as described in the previous section. This is followed by calculation cycles of what if scenarios, which follow a single-point deterministic modelling approach.

5.2.1 *Complexity of Cradle to Gate LCIA Results of Traction Batteries*

Model Setup

The objective of this section is to investigate the range of LCIA results for the production of battery cells due to their technological and spatial variability. As shown in Fig. 5.8, five different cathode chemistries are considered for the cathode; graphite is modeled as the anode. The production processes of the precursors required for the production of the cathode active materials are linked to the electricity mix of the

Chinese province of Hunan CN-HU, as a significant number of factories producing these materials are located in this region as reported in (Dai et al. 2018).

The created product objects are linked to five different spatial contexts, and correspondingly to six spatial contexts defining the background systems providing the electricity mixes for the manufacturing site. These are Germany (DE), Sweden (DE), France (FR), China-Guangdong (CN-GD), Japan (JP) and Korea (KR), motivated by the fact that these regions are not only intensively in the race towards building a battery cell production infrastructure, but also present strong variations among the electricity mixes available.

A total of 240 different product objects are created from the combination of the single-point values taken by the independent variables in the parameterized models. This led to a total of 1440 product systems after the product objects were linked with the previously mentioned product spatial contexts.

The numerical values of the parameterized variables at different levels of the integrated models in the ICLCE are shown in Table 5.5. The parameters for the case study are chosen for illustrative purposes. However, they represent some of the parameters that are heavily under discussion in the literature, as they are seen as an opportunity to decrease the manufacturing costs while increasing the battery pack's specific energy as discussed in (Petri et al. 2015; Wood et al. 2015; Ahmed et al. 2016; Schmuch et al. 2018; Cerdas et al. 2018b).

At the electrode level, the porosity of the cathode is parameterized in the model, and the single-point values chosen are assigned values ranging from 25 to 35%. This is consistent with experimental and simulation results seen in the literature, for example as reported in (Schmidt et al. 2020). At the stack level, the thickness of the coating of active material in the cathode is parameterized and assigned values ranging from 55 to 70 μm, which corresponds with the experimental and simulation

Table 5.5 Parameterized variables and single-point values given

Model level	Parameter	Single-point values	Unit
Cathode chemistry	Share of nickel (N), (Co) cobalt and manganese (Mn)	NMC 111, NMC 422, NMC532, NMC622, NMC811	–
Anode chemistry	Graphite (gr)	–	–
Electrode mass energy	Cathode porosity	0.25, 0.30, 0.35	% vol
Stack mass energy	Cathode tickness	55, 60, 65, 70	μm
Cell geometry	Cell size	173 × 125 × 45	mm
Cell geometry	Cell volume	0.97	L
Cell manufacturing	Dry room specific energy	30, 50, 70, 90	kWh/m^2 * day
Spatial context	Manufacturing site	JP, CN-GD, FR, SE, DE, KR	–
LCIA	Impact categories	GWP, TAP, FETP, MDP	–
LCIA	Characterisation model	Recipe	–

results reported in (Schmidt et al. 2020). Regarding manufacturing processes, two main variables are parameterized. First, the energy consumed by the dry room per square meter is parameterized based on the inputs provided by the description of the model given in (Schünemann 2015a; Ahmed et al. 2016).

Additionally, as described in the model coupling matrix, the cell manufacturing model is fed with product properties (e.g. thickness, porosity), which consequently influences the amount of energy required in processes such as cathode drying and the forming and aging stage of the process chain.

Results

The analysis of these results is strongly dependent on the impact assessment model selected, as the substances included in each of them, and the criteria followed for the definition of characterization factors, vary among the methodologies. The objective at this point is to demonstrate the advantages that the developed modelling concept in the context of this work might offer in enhancing the understanding of the complexity of the interactions between the different impact categories. Figure 5.9 shows the range of the calculated properties for all the product objects created. The *x-axis* shows the

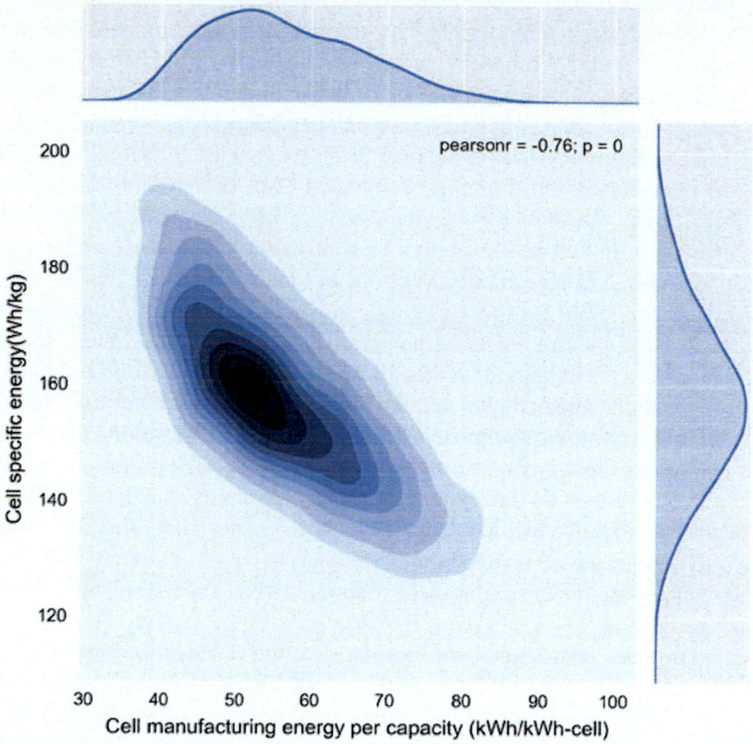

Fig. 5.9 Variation of cell specific energy and manufacturing energy demand per unit of energy capacity along the product objects created

spread of the total amount of energy demanded in the manufacturing stage, whereas the *y-axis* shows the spread of the specific energy calculated with the coupled models for all the product objects created. As seen, product properties at cell level, such as, in this case, the specific energy, variates in ranges between the 130 and 190 Wh/kg, which matches the current state of art specific energy densities as reported in (Placke et al. 2017; Schmuch et al. 2018). With the integrated model, manufacturing energy demand in the cell production stage was estimated to range between 40 and 80 kWh per unit of energy capacity, which is relatively low when compared to the results reported in the LCA literature summarized in Table 3.3. The results are nevertheless fairly comparable to the reported values in more current research done by (Davidsson Kurland 2019). The authors followed a top down approach, which estimated the energy consumption at the *Tesla Gigafactory 1* and the *Northvolt* cell factory in Skelleftea to range between 50 and 65 kWh per unit of kWh of cell capacity. A trivial but nevertheless relevant trend to notice is the correlation between specific energy and the manufacturing energy.

As reported in different scientific publications, increasing energy density reduces energy footprint per unit of cell capacity. This strategy can lead to different effects depending on the electricity mix supplied at the manufacturing site. Also, this can variate depending on the environmental impact category analyzed, as shown in the results further in this section. The cradle to gate environmental impacts were estimated for a 50 kWh battery pack built with each of the battery systems created. A contribution analysis was then performed to understand the impact of manufacturing energy on the total supply chain of a battery pack.

The results are shown in Fig. 5.10, where the contribution is plotted against the specific amount of energy demanded. Two observations are worth mentioning here. First, the higher the carbon content of the electricity mix used during the manufacturing phase, the greater its contribution to the total cradle to gate footprint.

Second, a higher energy footprint per unit of energy capacity is not significant in terms of GWP for those battery cells produced in manufacturing sites with electricity mixes with a lower carbon content. In these cases, other factors (such as the production of the cathode materials; the production of the current collectors required for the anode and cathode; and the production of the solvent used for the production of the slurry mixture) contribute in almost equal shares to the C2G GWP.

These observations are reflected in Fig. 5.11. The figure shows the variability of the GWP results throughout the product systems as a density plot clustering the different manufacturing sites for these three impact categories. The results in the case of GWP are spread from 65 kg CO_2-eq/kWh to approximately 180 kg CO_2-eq/kWh, which are relatively similar to those summarized from the scientific literature in Table 3.2.

The figure also shows, for instance, that the GWP of the battery cells produced in places like Sweden and France are less sensitive to the spread of the energy footprint than, for instance, battery cells manufactured in the province of Guangdong China. Moreover, as seen in the figure, the curve of the battery cells produced in Sweden is displaced to the left side of the of the chart. The difference between the GWP results of the battery cells produced in these places can range from 35 to 80 kg CO_2-eq/kWh.

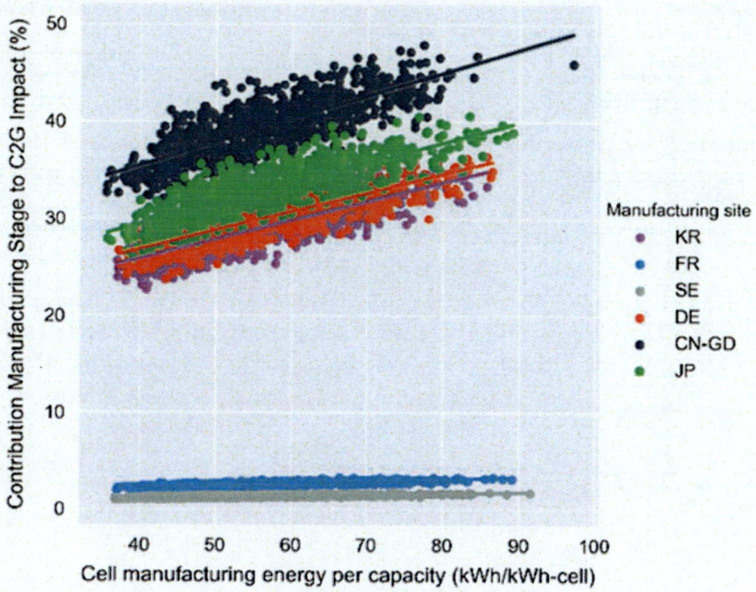

Fig. 5.10 Variation of the contribution of the manufacturing energy to the C2G GWP for different manufacturing sites. KR: Korea, FR: France, SE: Sweden, DE: Germany, CN-GD: China Guangdong province, JP: Japan

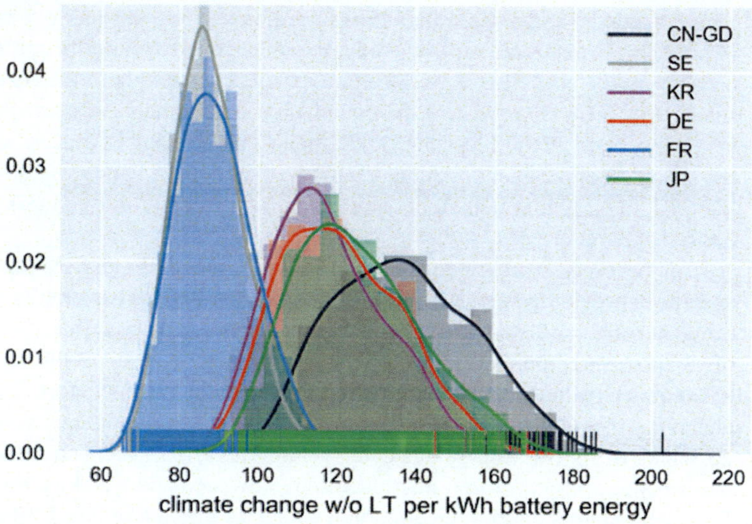

Fig. 5.11 Variability of GWP results per kwh of battery energy for the different manufacturing sites considered

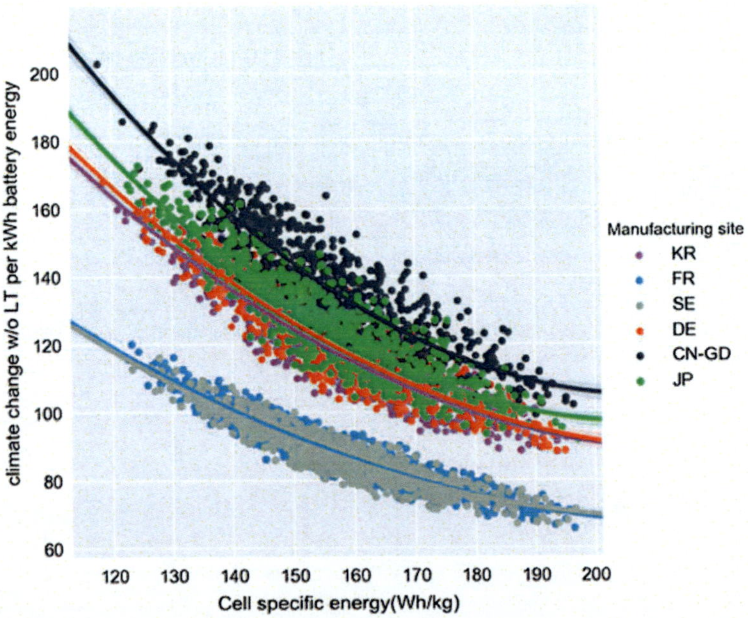

Fig. 5.12 Effect of increasing specific energy at cell level to the GWP per unit of energy capacity

As discussed and represented in Fig. 5.12, increasing the energy content per unit of mass in a battery cell implies a reduction in the energy footprint per unit of energy capacity in the cell. This is, to a certain extent, transferable to the case of GWP, as seen in Fig. 5.14. Disregarding the manufacturing site, increasing gravimetric energy at cell level not only reduces the GWP impact unit of energy capacity, but also limits the influence of the energy demanded by the manufacturing processes as the cathode material, in particular nickel, becomes the main contributor to the C2G GWP of the battery system.

This becomes more evident in Fig. 5.13a, b, where the correlation between the specific energy of the cell and the GWP per energy capacity is plotted while considering the effect of the cathode chemistry. Notice that battery cells with a higher share of nickel are predominantly located in the lower left quadrant of the chart. The influence exerted by the amount of nickel is, as given in the graphic, more significant if the electricity mix of the manufacturing site has a higher carbon intensity. An increase of 20 Wh/kg of specific energy at a cell level implies a 10 kg CO_2-eq reduction per unit of energy capacity for a cell manufactured in Sweden and a 20 kg CO_2-eq reduction for a cell manufactured in Guangdong, China, as implied in Fig. 5.13.

While Fig. 5.13 suggests that increasing the cell's specific energy by increasing the quantity of nickel in the cathode mix potentially decreases GWP footprint per unit of energy capacity, the effect of increasing the share of nickel might potentially have a negative effect on other environmental impact categories. One of the advantages of having access to LCIA results of such a large number of product systems is the

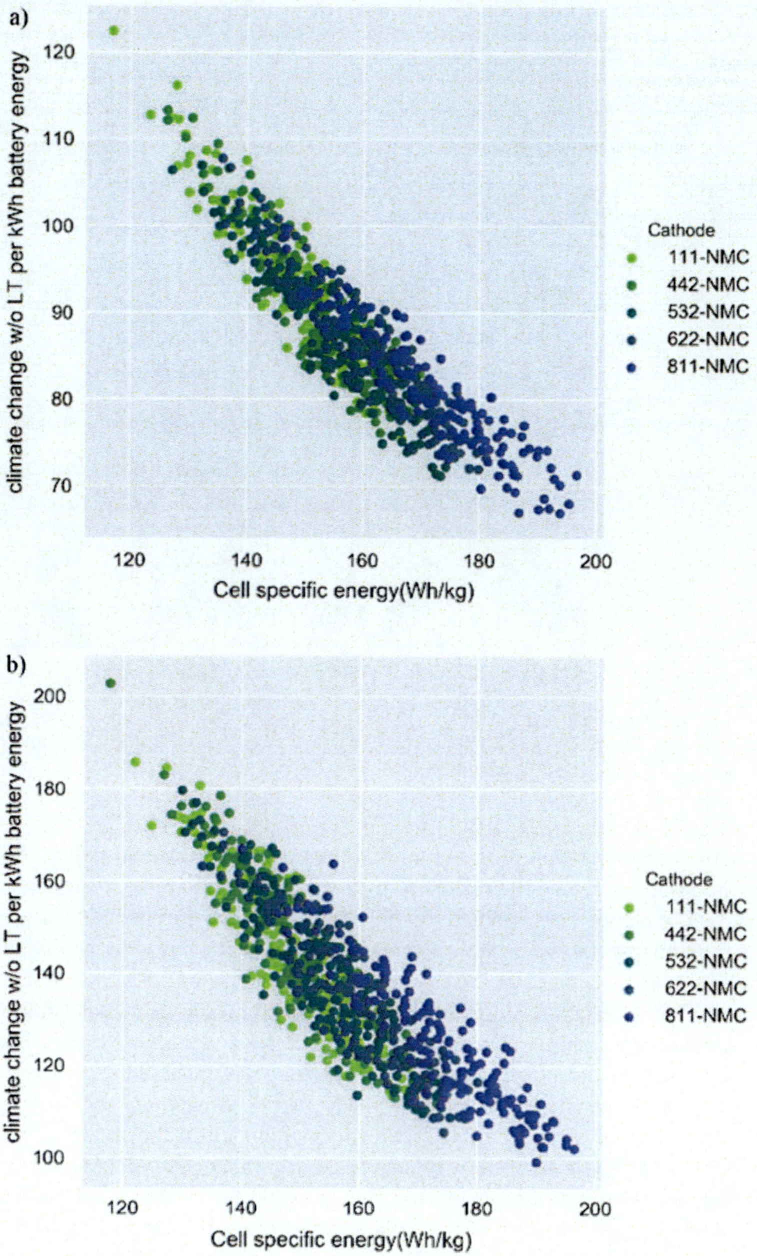

Fig. 5.13 Effect of increasing specific energy at cell level to the GWP per unit of energy capacity and the effect of the cathode chemistry. **a** Manufacturing site in Sweden, **b** Manufacturing site in Guangdong, China

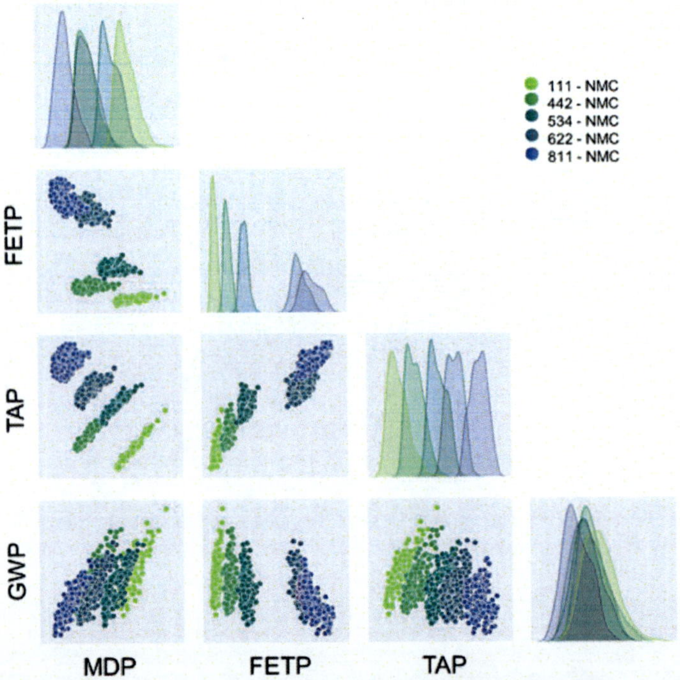

Fig. 5.14 Pairplot of C2G LCIA for all the product systems evaluated. GWP: Global Warming Potential, FETP: Freshwater Ecotoxicity Potential, TAP: Terrestrial Acidification Potential, MDP: Mineral Depletion Potential

possibility of developing support to better understand the so-called problem shifts or trade-offs between the different impact categories being investigated.

A pairplot is presented in Fig. 5.14 for this purpose. It shows pairwise the relationship between the different impact categories results that are exemplary analyzed.

The figure shows a grid of axes, which enables a visual association between the impact categories in order to discover synergies and possible trade-offs. The diagonal shows the distribution plots of each of the impacts clustered in groups for the different cathode chemistries considered in the model. Several observations regarding the tradeoffs between the impact categories included in the model can be made from Fig. 5.14. These are mostly related to the significance of nickel to the toxicity related impact categories arranged in the plot. Nickel is the second most significant substance characterized in most of the toxicity related environmental impacts in the Recipe methodology, as reported by Huijbregts and colleagues (Huijbregts et al. 2016). In general, it can be notice that gains in GWP are independent from the influence of the product systems to the other three impact categories. Interesting interactions can be seen, for instance, between TAP, MDP and FETP, in which an increase in the share of nickel per unit of active material mass represents an increase in the environmental impact in terms of TAP and FETP, and a decrease in the environmental impact in terms of MDP, as the share of manganese per unit of active material mass increases.

In this regard, the discussion in this part is essentially focused on highlighting the potentials of the results of an ICLCE approach for traction batteries in supporting decision making in engineering. In the case of traction batteries, two mitigation strategies can be discussed towards the decarbonization of battery cells. First, the decarbonization of the energy required during the manufacturing of battery cells (background system), and the increase of the energy efficiency (foreground system) plays a key role. As reported by Michaelis and colleagues (2018) before the year 2030 not only is a strong increase in the production capacities of current cell manufacturers expected, but also a major shift of the production sites. Particularly driven by the interest of producing close to the demand hotspot, Japanese and Korean manufacturers have increasingly strengthened their plans to shift production to Europe and U.S.A. Considering aspects such as the carbon intensity of electricity during a decision process regarding the manufacturing site can result in a significant reduction of the cradle to gate GWP of traction batteries, and therefore of the electromobility sector.

As observed in Fig. 5.15, a good mitigation potential for the shifting of production sites from countries such as China, Korea and Japan to places with a lower carbon intensity in the local electricity mix. Decarbonization through improvements in the foreground system can be mostly achieved through increasing energy efficiency (e.g. through optimizing processes or increasing production volumes). Additionally, by developing tailored strategies to diversify energy carriers supplying energy to particularly relevant consumers, such as the machines required for the dry-room, aging and forming and electrode drying processes (districts heating and cooling link in the city).

Increasing specific energy of battery cells represents a significant strategy towards decarbonizing the production of battery cells. For this reason, design strategies such as increasing the thickness of the cathode or decreasing the porosity of the active material through an enhanced processing can lead to a reduction of different reduction values, depending on the manufacturing site. So, an increase of 40 Wh/kg in

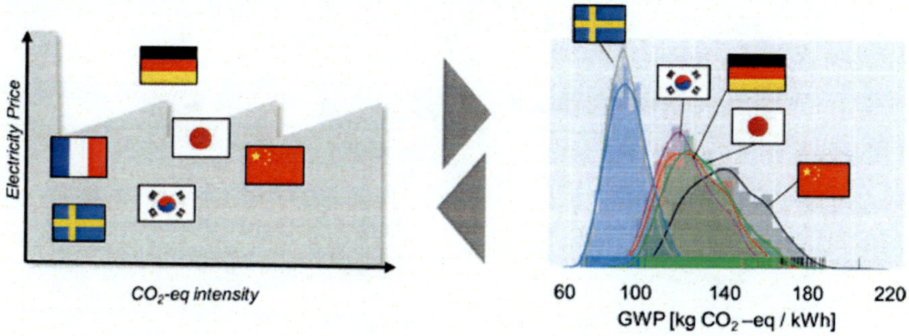

Fig. 5.15 Variability of the specific GWP impact per km [gCO_2-eq/km]

the cell specific energy represents a reduction of around 25 kg CO_2-eq per kWh of energy capacity for a factory located in Sweden and 45 kg CO_2-eq per kWh for a cell produced in a factory in China. This is a valuable insight, as it allows for the prioritization of decarbonization strategies regarding product and production parameters. Increasing specific energy in battery cells can be achieved through optimizing cell geometries and through material selection and development. For example, cell chemistries with a higher Ni content are estimated to offer a higher energy density. Nevertheless, increasing the amount of Ni would imply a larger share of the environmental impact of a battery cell to be material dependent which, in the case of Ni, implies a raise in the environmental issues in toxicity related impact categories.

5.2.2 *Complexity of LCIA Results of EVs Usage Stage*

Model Setup

The second part of the case study focuses on demonstrating how the ICLCE approach can be implemented to support the analysis of the variability of EV LCIA results due to their technological and spatial discrepancies during the usage stage. The approach followed in this section is shown in Fig. 5.16. In this part, the model described in the Eqs. 5.18–5.25 is parameterized so as to be able to calculate the mass and

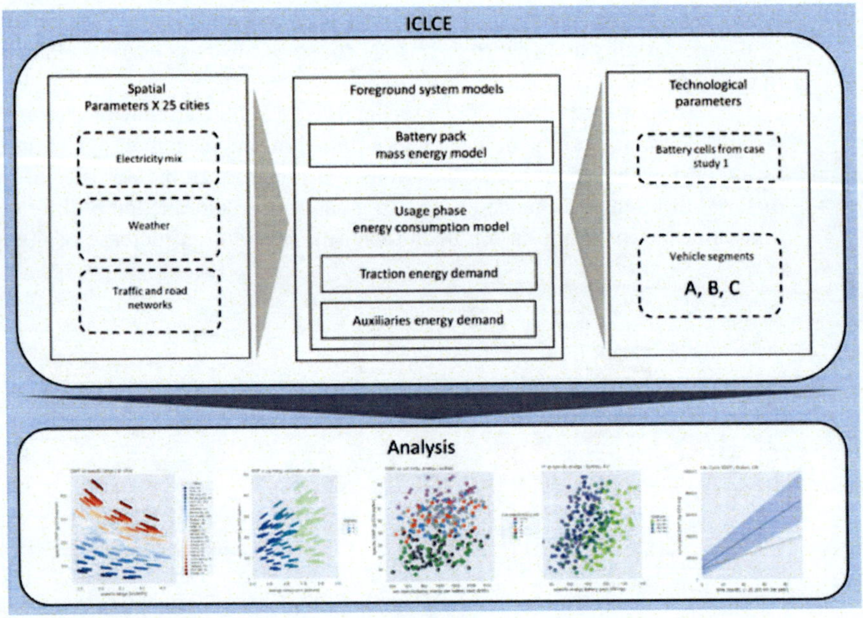

Fig. 5.16 Methodology followed for the case study #2

Table 5.6 Geometrical properties of the vehicles considered in the case study

Segment	Curve weight (kg)	Battery pack active volume (l)	A*Cd (m^2)
A	900	55.80	0.600
B	1139	67.32	0.644
C	1510	87.32	0.750

energy capacity of different vehicle geometries, combined with the previously defined battery cell objects.

In this regard, 36,000 new product objects are created considering the vehicle's information summarized in Table 5.6. Here, information regarding weight, cell manufacturer and number of cells is compiled for each of the European class segments. Further, the vehicles construction space allocated for battery cells is calculated as shown in Table 5.6.

Further, each defined product object (i.e. vehicle equipped with a traction battery system) is linked to 25 different spatial contexts where the usage stage is assumed to take place. In each of the cities, the electricity mix was considered to be fixed as given in the ecoinvent database. The climate temperatures are considered to vary hourly and throughout the year as described in Sect. 5.1.2.

Spatially explicit traffic information from data providers such as open route service ORS are consolidated for each of the cities and classified into morning peak traffic, evening peak traffic, free flow traffic and average flow traffic. For this purpose, a bounding box is programmatically placed as a border in each of the cities where the street network ends. Defined number coordinate points (in this case 1000) are programmatically and randomly located inside the bounding box for the different cities considered. Routes are generated by randomly defining the start and end coordinates (see Fig. 5.17). Through an API to the ORS services, information regarding average speed and speed segments for each of the routes is retrieved and consolidated. This information is further coupled with consolidated real time traffic information reporting on traffic congestion in cities such as the data provided in the TomTom traffic index (Tomtom 2018). In this way, it is possible to obtain speed profiles for each city for different moments of the day. The segments of the routes are linked to particular segments of the WLTP driving cycle as summarized in Table 5.4, enabling the approximation of the energy demanded for each route.

Next, a weekly driving profile mix was defined by assuming particular shares of types of trips during a week, as shown in Table 5.7. Under this assumption, a total yearly driven distance of 20,280 km was estimated. The total lifetime of the vehicles was defined to be 200,000 km.

Results

The analysis is focused exclusively on GWP. Figure 5.18 illustrates the spread of the speeds for each of the routes modelled for the city of Warsaw. Notice that in terms of probability density of the distribution of the speeds, the curves depicting morning and evening traffic peaks are displaced to the left compared to the distribution curve

Fig. 5.17 Random defined routes inside the city's bounding box for Warsaw, Poland. Plotted with Leaf© and OpenStreetMaps© (OpenStreetMap contributors 2015)

Table 5.7 Assumed weekly driving profile

Profile	km/week
Morning commuting distance	150
Evening commuting distance	150
Free flow trips distance	40
Average flow distance	50

of the average flow of speed of the routes. This slight reduction in the speed affects the traction energy consumption, as the segment is then linked to a segment of the driving cycle with lower average speed. It also affects the amount of energy consumed by the auxiliary devices in the vehicle, as the duration of the trip increases. This information is shown for all the cities considered in Appendix. Figure 5.19 shows the yearly outside temperature profile per hour for the city of Warsaw, Poland (for illustrative purposes). The temperature profiles for all the cities considered in this part of the case study are shown in the appendix at the end of this chapter.

Figure 5.20 presents exemplarily the results of coupling the results of the models presented in Figs. 5.18 and 5.19 with the technical properties for a vehicle of the B segment. The figure presents the electricity consumption mix per month for that specific vehicle for each month for the city of Warsaw, Poland (just as an example. The results of the rest of the cities can be seen in the appendix). The results for the same vehicle in all of the cities considered in the model are shown in the appendix at

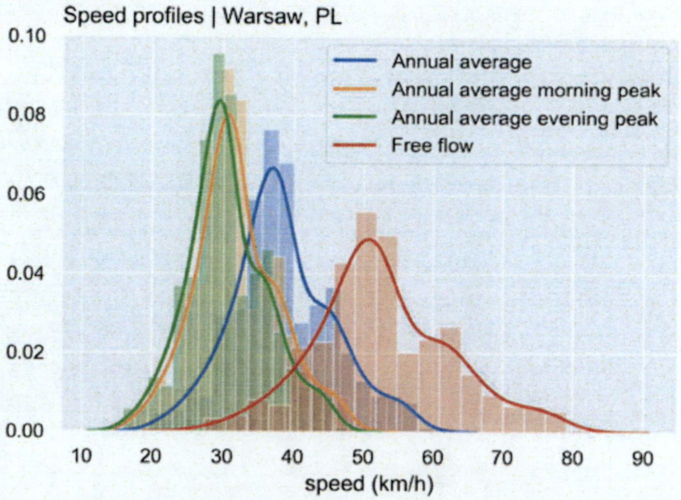

Fig. 5.18 Consolidated speed profiles for Warsaw, Poland

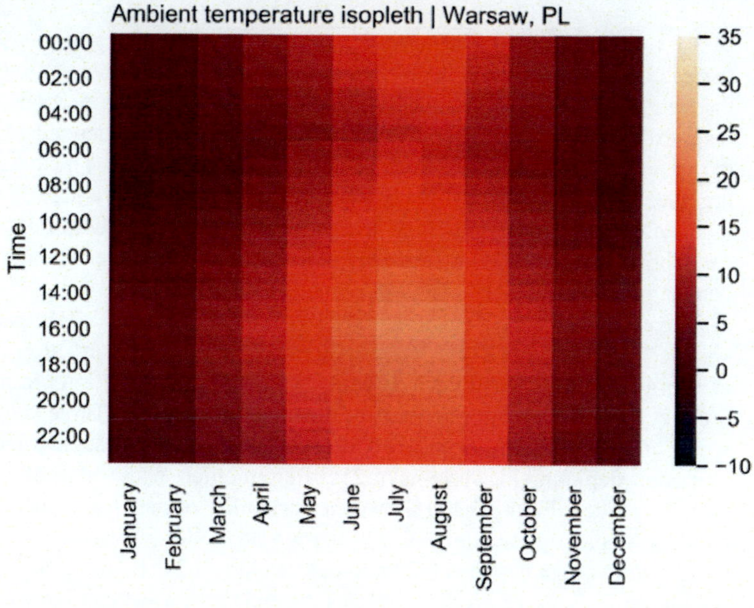

Fig. 5.19 Consolidated hourly temperature profile for Warsaw, Poland

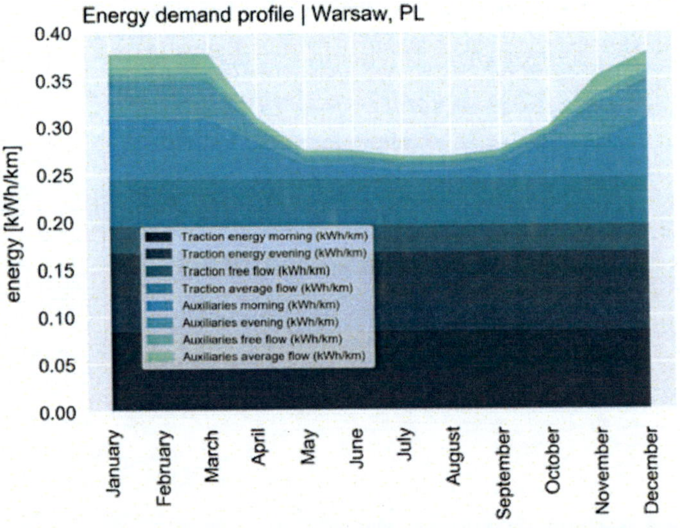

Fig. 5.20 Consolidated energy consumption mix profile for Warsaw, Poland

the end of this chapter. As seen, the energy mix consist of four static (no variations throughout the year) and four dynamic sections. The static sections are basically responsible for the technological variability, as it is mainly dependent on the speed of the vehicle throughout the route and its mass and geometry. The dynamic sections, together with the electricity mix used, will affect the spatial variability of the results, as they are mainly dependent on temperature and trip duration. For this particular case shown in Fig. 5.21, a difference of almost 30% of the average energy consumed per km is estimated.

The effects of technological variability on the life cycle GWP impacts of electric vehicles are plotted in Fig. 5.21. This figure presents the interaction of the amount of energy required per km driven with the GWP impact per km within each of the product systems operating in each of the cities considered. Three vehicle segments were equipped with a total of 1000 different battery cells product systems and linked to models depicting the systems of 25 different cities, with a total of 75,000 product systems plotted. Figure 5.21a shows the variability of the results due to the variation of technological parameters. While the average energy consumption per km was estimated to range between 0.23 and 0.42 kWh (~ 82% higher), the life cycle GWP per km throughout the product systems created was estimated to range between ~ 50 and 450 gCO_2-eq/km (~ 900% higher). Some observations stand out. First, the size and mass of the vehicle obviously has a significant effect on the energy demanded per km. Nevertheless, the influence on the GWP impacts per km have different dimensions depending on the spatial contexts where the vehicles are operating, as seen in Fig. 5.21b. Vehicles of the segment A are estimated to achieve values spread between ~ 50 and 300 gCO_2-eq/km, vehicles in segment B between ~ 60 and 375 gCO_2-eq/km, and vehicles in segment C ~ 75 and 450 gCO_2-eq/km. This wide

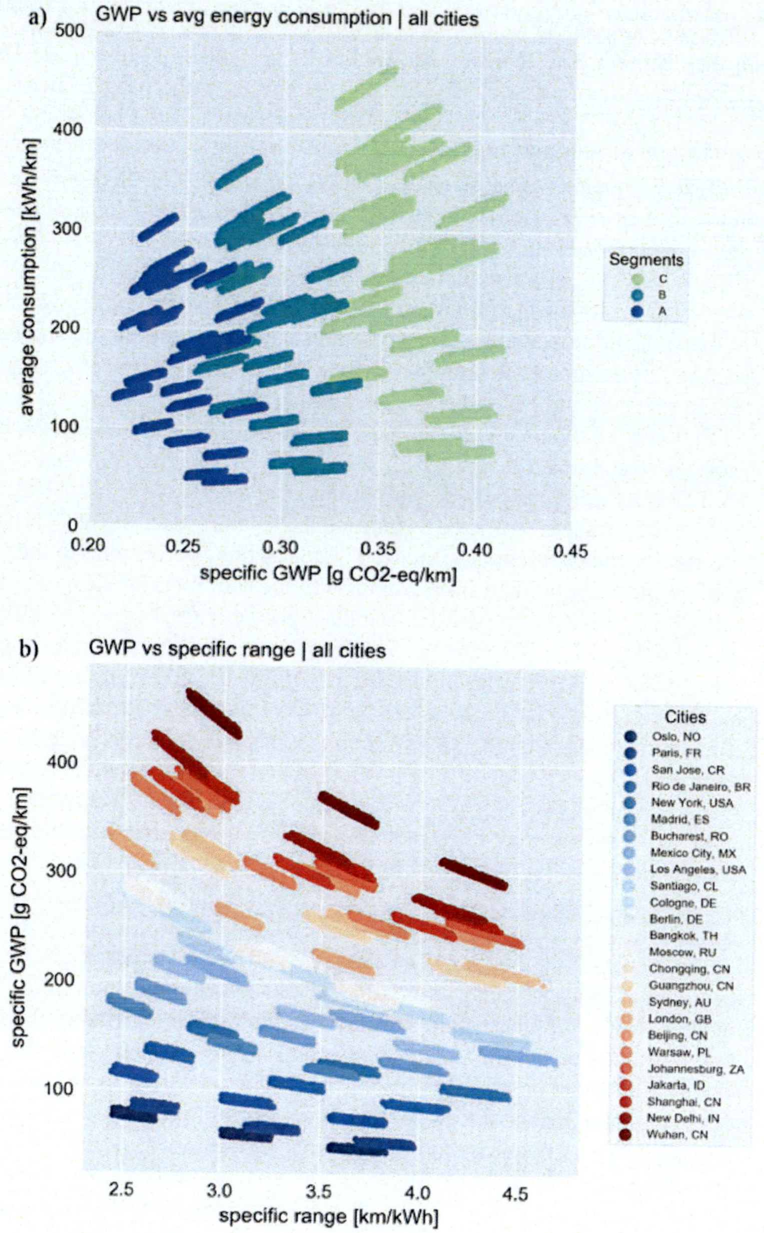

Fig. 5.21 Variability of the specific GWP impact per km [gCO_2-eq/km] due to technical (**a**) and spatial variability (**b**)

difference among the technologies is, unsurprisingly, driven by the electricity mixes in the different cities considered. This turns up scaled in those cities with extreme temperature seasons, especially in those countries with cold weather conditions in winter. In Fig. 5.21b, around half of the cities are located below the limit of 200 gCO_2-eq/km (points in blue shades).

In this segment, some places such as Oslo, Norway or San José, Costa Rica are among the cities offering the best potential reduction of GWP impacts per km because the local electricity mixes in those places contain almost completely renewable energies. Paris, France, is also included in this blue segment, due to their strong reliance on nuclear energy, which is not reflected in GWP impacts. Places like Wuhan, China or Jakarta, Indonesia are among the cities in which not only the amount of energy required by auxiliaries is high, but also the electricity mix is characterized by having a high share of fossil-based energy sources.

Figure 5.22 shows break even plots for the cities of Sydney, Australia and Oslo, Norway. Break-even diagrams for all the cities considered in this case study are included in the appendix at the end of this chapter.

The blue shade is estimated by plotting the results at each km of all product systems created for the different case study. The black lines correspond to the ICE vehicles and are plotted with data from real driving measurements done by the international council on clean transportation, which reported a total of ~ 155 gCO_2eq/km for diesel and gasoline for vehicles in Europe in 2019 (International council on clean transportation 2019). The embodied variability of GWP impacts brought by the battery cells to the usage stage can be observed in Fig. 5.22, at kilometer 0 on the *x axis*. Notice the cradle to gate GWP impact of all the electric vehicles vary between the 5 and 20 ton CO_2-eq. While this difference remains steady in places with 100% renewable energy sources (Fig. 5.22b), in places where the electricity mix and other influencing factors play a key role, this difference increases with the driven distance among the product systems (Fig. 5.22a, c, d). At the kilometer 200,000, the variability of the results is enlarged. As seen, in the case of Sydney, the range of the results goes from 35 to 70 ton CO_2-eq.

As argued at the beginning of this chapter, the objective of the case study goes beyond the analysis of the results achieved after the application of the implemented models. Rather, the case study fulfills the goal of demonstrating the powerful and unconstrained modelling and analysis possibilities enabled by the ICLCE concept. The results of the case study didn't necessarily disclose new insights regarding the environmental impact of EVs compared to the reported results in the literature or the general current knowledge on the environmental impact of EVs. Arguing whether a technology is environmentally better is not only misleading, as there is a huge knowledge gap in upstream life cycle stages such as the material production phases, but also misses the point of the concept presented in this thesis. For the question "*are EVs environmentally better than ICE vehicles?*" the answer "*it depends*" not only remains valid, but becomes more significant as it can contribute toward guiding the improvement of products and systems. With an ICLCE modelling approach, the scope of LCE goes beyond focusing on finding a single result (e.g. x g CO_2-eq/km) to starts reflecting on the circumstances that cause the results to vary, as well as on

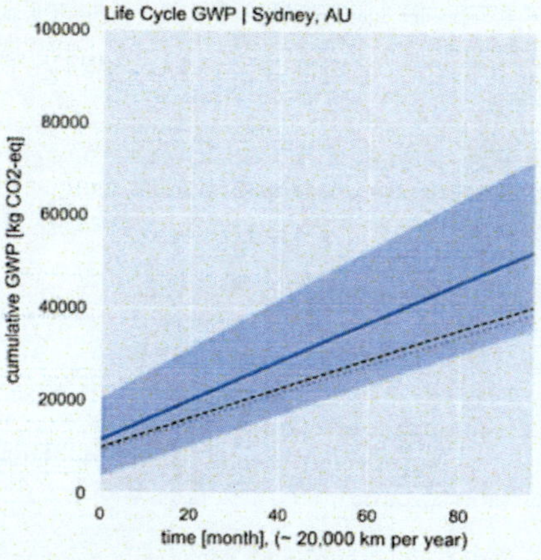

(a)

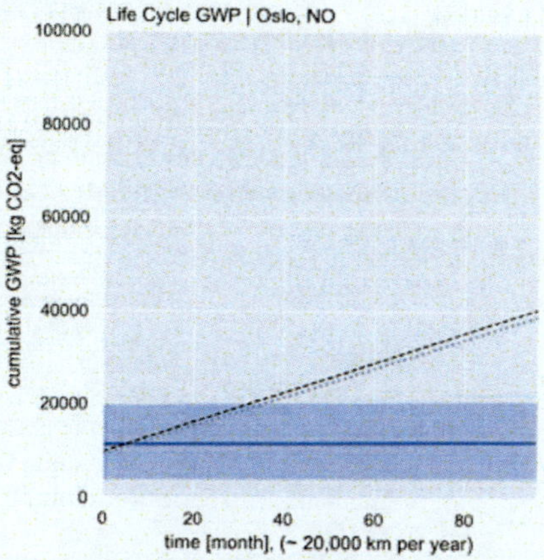

(b)

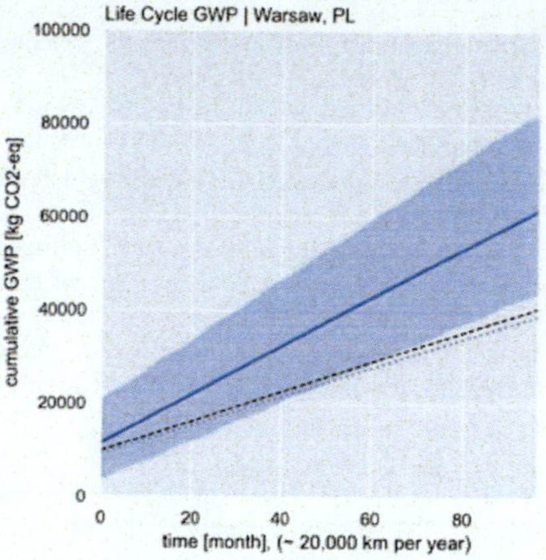

(c)

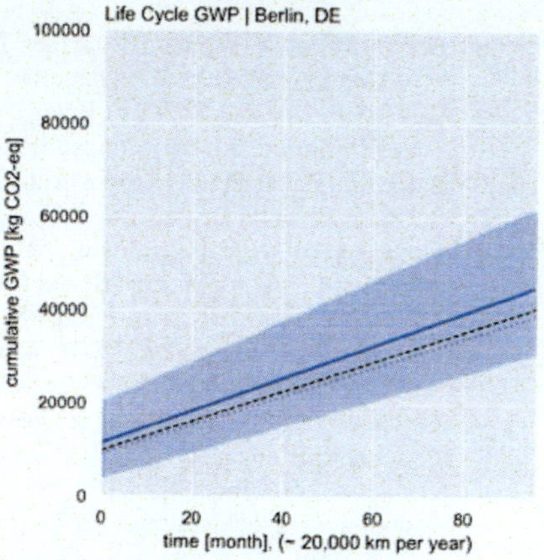

(d)

Fig. 5.22 Robust comparative break-even analysis between EV and ICE in **a** Sydney AU, **b** Oslo, NO, **c** Warsaw, PL, **d** Berlin, DE. The blue shade represents all possible results for EVs. The black dotted lines represent ICE vehicles

the correlations between technical properties and environmental impacts, and the trade-offs between different solutions and systems.

References

Ahmed S, Nelson PA, Dees DW (2016) Study of a dry room in a battery manufacturing plant using a process model. J Power Sources 326:490–497. https://doi.org/10.1016/j.jpowsour.2016.06.107

Beck HE, Zimmermann NE, Mcvicar TR et al (2018) Data descriptor: present and future Köppen-Geiger climate classification maps at 1-km resolution. Nat Publ Gr 5:1–12. https://doi.org/10.1038/sdata.2018.214

Betz J, Bieker G, Meister P et al (2019) Theoretical versus practical energy: a plea for more transparency in the energy calculation of different rechargeable battery systems. Adv Energy Mater 9:1–18. https://doi.org/10.1002/aenm.201803170

Carriere D (2018) Geocoder: simple, consistent

Cerdas F, Titscher P, Bognar N et al (2018) Exploring the effect of increased energy density on the environmental impacts of traction batteries: a comparison of energy optimized lithium-ion and lithium-sulfur batteries for mobility applications. Energies 11:150. https://doi.org/10.3390/en11010150

Cerdas F, Gerbers R, Andrew S et al (2018a) Disassembly planning and assessment of automation potentials for lithium-ion batteries. In: Recycling of lithium-ion batteries the LithoRec way, pp 83–97

Dai Q, Kelly CJ, Dunn J, Benavides TP (2018) Update of bill-of-materials and cathode materials production for lithium-ion batteries in the GREET Model. US Dep Energy

Davidsson Kurland S (2019) Energy use for GWh-scale lithium-ion battery production. Environ Res Commun 2:012001. https://doi.org/10.1088/2515-7620/ab5e1e

Duce AD, Egede P, Öhlschläger G et al (2013) Guidelines for the LCA of electric vehicles. European Union Seventh, Brussels

Egede P (2017) Electric vehicles, lightweight design and environmental impacts. Springer, pp 9–40

Ellingsen LA-WW, Majeau-Bettez G, Singh B et al (2014) Life cycle assessment of a lithium-ion battery vehicle pack. J Ind Ecol 18:113–124. https://doi.org/10.1111/jiec.12072

Gao DW, Mi C, Emadi A (2007) Modeling and simulation of electric and hybrid vehicles. Proc IEEE 95:729–745. https://doi.org/10.1109/JPROC.2006.890127

GIScience (2008) OpenRouteService

Graphhopper contributors (2018) Graphhopper

Grunditz E, Thiringer T (2016) Performance analysis of current BEVs—based on a comprehensive review of specifications. IEEE Trans Transp Electric 7782:1–1. https://doi.org/10.1109/TTE.2016.2571783

Hawkins TR, Singh B, Majeau-Bettez G, Strømman AH (2013) Comparative environmental life cycle assessment of conventional and electric vehicles. J Ind Ecol 17:53–64. https://doi.org/10.1111/j.1530-9290.2012.00532.x

Hofer J (2014a) Sustainability assessment of passenger vehicles: analysis of past trends and future impacts of electric powertrains. research-collection.ethz.ch

Hofer J (2014b) Sustainability assessment of passenger vehicles: analysis of past trends and future impacts of electric powertrains. ETH Zurich

Huijbregts M, Steinmann ZJN, Elshout PMFM et al (2016) ReCiPe 2016. Natl Inst Public Heal Environ 194. https://doi.org/10.1007/s11367-016-1246-y

International council on clean transportation (2019) fact sheet: Europe gasoline vs. diesel comparing Co_2 emission levels of a modern medium size car model under laboratory and on-road testing conditions. theicct.org

Kottek M, Grieser J, Beck C et al (2006) World Map of the Köppen-Geiger climate classification updated. Meteorol Zeitschrift 15:259–263. https://doi.org/10.1127/0941-2948/2006/0130

Kwade A, Diekmann J, Hanisch C et al (2016) Recycling von Lithium-Ionen-Batterien—LithoRec II Abschlussberichte der beteiligten Verbundpartner. Verbundprojekt im Rahmen des Förderprogramms „Erneuerbar Mobil“ des Bundesministeriums für Umwelt, Naturschutz, Bau und Reakt

Michaelis S, Rahimzei E, Kampker A et al (2018) Roadmap Batterie-Produktionsmittel 2030. VDMA

Notter DA, Gauch M, Widmer R et al (2010) Contribution of Li-Ion batteries to the environmental impact of electric vehicles. Environ Sci Technol 44:6550–6556. https://doi.org/10.1021/es903729a

OpenStreetMap contributors (2015) OpenStreetmap Planet dump retrieved from https://planet.osm.org

Petri R, Giebel T, Zhang B et al (2015) Material cost model for innovative li-ion battery cells in electric vehicle applications. Int J Precis Eng Manuf Technol 2:263–268. https://doi.org/10.1007/s40684-015-0031-x

Placke T, Kloepsch R, Dühnen S, Winter M (2017) Lithium ion, lithium metal, and alternative rechargeable battery technologies: the odyssey for high energy density. J Solid State Electrochem 21:1939–1964. https://doi.org/10.1007/s10008-017-3610-7

Schmidt O, Thomitzek M, Röder F et al (2020) Modeling the impact of manufacturing uncertainties on lithium-ion batteries. J Electrochem Soc 167:060501. https://doi.org/10.1149/1945-7111/ab798a

Schmuch R, Wagner R, Hörpel G et al (2018) Performance and cost of materials for lithium-based rechargeable automotive batteries. Nat Energy 3:267–278. https://doi.org/10.1038/s41560-018-0107-2

Schünemann J-H (2015a) Modell zur Bewertun der Herstellkosten von Lithiumionenbatteriezellen

Schünemann JH (2015b) Modell zur Bewertung der Herstellkosten von Lithiumionenbatteriezellen. Sierke

Tomtom (2018) Traffic Index—Traffic results 2018. 21

Wood DL, Li J, Daniel C (2015) Prospects for reducing the processing cost of lithium ion batteries. J Power Sources 275:234–242. https://doi.org/10.1016/j.jpowsour.2014.11.019

Chapter 6
Summary, Critical Review and Outlook

Contents

This conclusion presents a summary of the content in each of the chapters, in addition to a critical reflection on the developed methodology. An outlook is provided at the end, giving a summary of possible further research directions.

6.1 Summary

The concept presented in this book contributes towards advancing research and development in the field of the sustainability assessment and life cycle engineering of complex product systems. Particularly, the work provides solid foundations for the development of computational tools to enable the seamless integration of LCA within engineering activities, while at the same time conserving the integrity and comprehensiveness of the methodology and without falling into extreme simplifications. The research on future mobility urgently requires such methodological developments and their resulting tools. Chapter 1 in this thesis gives an overview of the current state of the field of traction batteries for EVs, and outlines the partially controversial discussion regarding the environmental impacts related to electromobility. The chapter goes on to discusses the necessity of new tools and methodologies to support the development of new technologies in this field.

Chapter 2 introduces the theoretical background in which this thesis has been framed. It is structured into three main parts. First, an overview of the field of environmental sustainability and LCE is introduced. Here, the strengths and weaknesses of the LCA methodology and its application in engineering is discussed. Second, an overview of the technical properties of traction batteries and electric vehicles is

F. Cerdas, *Integrated Computational Life Cycle Engineering for Traction Batteries*, Sustainable Production, Life Cycle Engineering and Management,
https://doi.org/10.1007/978-3-030-82934-6_6

given. Finally, the interface of both topics is described at the end of the chapter. A special focus has been given to the issue of complexity and variability when applying LCA to the analysis of electromobility, and to the challenges faced during decision making processes in this regard.

An overview of the current relevant research surrounding and supporting this thesis is provided and analyzed in Chap. 3. The collection and analysis of approaches was structured into three groups: (i) environmental assessment studies in the field of electromobility and traction batteries, (ii) (integrated) computational approaches in the field of electromobility and traction batteries and (iii) (integrated) computational approaches in adjacent engineering fields focused on the modelling of environmental implications of technological systems. The research gap identified is further explained at the end of the chapter.

Chapter 4 starts with a synthesis of the requirements for an integrated computational life cycle engineering approach for EV. The analysis builds upon the identified research gap presented in Chap. 3. The objectives are then defined and the methodology is introduced, beginning with a generic framework for an Integrated Computational Life Cycle Engineering. Next, a reference architecture for the development of an ICLCE model platform for EV and traction batteries is introduced, and the respective computational layers are then presented. Finally, an implementation concept describing the main information flows and keys elements of the system is given.

Chapter 5 has the objective of illustrating the construction of an ICLCE model for the analysis of Evs and traction batteries, and of exemplifying how the application of such a model can contribute towards enhancing the understanding of the variability and uncertainty in the results of environmental impact in this field. For this purpose, a model coupling strategy is introduced at the beginning of the chapter, including a system definition and the identification of the models required to be integrated. Next, the key foreground system models from relevant disciplines are described, as well as the different spatial and background system models and/or datasets linked to the foreground system. Finally, a case study is presented at the end of the chapter. It consists of two sections: (i) an analysis of the variability and complexity of the cradle to gate environmental impacts of traction batteries and (ii) an evaluation of the usage stage of electric vehicles with an enhanced spatial and temporal variability, while considering the important influencing factors that drive environmental consequences in the usage stage.

6.2 Critical Review

The developed concept offers a unique approach to create LCE models for complex product systems. It is based on the coupling of models from different disciplines within an integrated platform, which enables the creation of product systems that consider all possible variations raising from model parameters and/or models assumptions. By doing this, an LCE model is not seen any longer as a box of inputs and outputs linking technical assumptions made by scientific experts to represent a certain context. Rather, it becomes a connected *system model*, composed of stand-alone models that deliver the necessary information to models in the up and down-stream life cycle stages. This results in consistent product systems based on models outputs and interactions rather than on disconnected assumptions.

ICLCE has been partially inspired by the ICME community, which is focused on unravelling the path from materials to product properties (Horstemeyer and Sahay 2018). ICME's overall goal is to understand how, for specific material properties, a particular set of processing steps will change the structure of the materials being analyzed. These changes in the structure affect or create material properties that are desired for a particular application. The link between material properties, processes, structures and product performance is only possible thanks to an organized and harmonized group of models depicting individual parts of the system, while at the same time providing the necessary information to understand the performance of the whole.

The whole in LCE is a product system, and its performance is its interaction with the biosphere in terms of the consequences driven by the type, properties and amount of elementary flows crossing the border between the manmade system and nature. In this regard, ICLCE is as well inspired on the Lyngby framework for LCE (Hauschild et al. 2017), which encourages shifting the focus of LCE towards addressing the environmental dimension of sustainability, and proposes product systems to be *life cycle engineered* while considering as high as possible a resolution of the influencing factor at the different technical, spatial and socioeconomic layers through which a product system is spread.

Following a two-ways modelling paradigm (i.e. from design through product life cycle, life cycle inventory to Impact assessment) embedded in a W-model to engineer product systems, the ICLCE concept in this thesis enables considering the inherent variability in the composition of a LCI of a product system, while enabling an assessment of the effects linked to these emissions.

The concept was developed particularly to support the coming research in the field of traction batteries for electric vehicles. As the transportation sector transitions towards its electrification, the products (e.g. vehicles and components), the processes (i.e. manufacturing processes and material supply chains) and the business models (i.e. connectivity, sharing schemes, second life, leasing and so on) have already started to suffer strong transformations. This will lead to an unprecedented increase in the customization of products, and therefore in its complexity, making current LCA approaches obsolete. Current LCAs on electromobility often look into the past of a generic product system using generic average data that is not necessarily representative. One generic dataset for a car-body of an ICE is linked to a LCI of a battery system modelled as a separated block, partly with assumptions gathered from interviews or non-validated laboratory data, and the results are intended to support decision making concerning future Evs. Future LCAs on electromobility will be able to represent the real EV as its architecture changes, while synergistically integrating more functions, becoming more connected and while its function within an urban mobility system increasingly reshapes. This will only be possible with large scale system models.

It is not yet within the scope of this book to go deeper into the meaning behind the results, in terms of the real consequences and implications of traction batteries to the environment. This discussion, while necessary, requires another frame in which the elementary flows of the product systems created are analyzed against the models and assumptions behind the different impact categories. This means, for instance, considering in detail the substances included in the characterization matrixes of the impact assessment methodologies, the compartments and geographical regions where the elementary flows are assumed to be emitted.

In this way, the case study presented in this study has the function of exemplifying how, by following the modelling concept, one or many LCE models can be created out of coupling discipline specific models. Several assumptions regarding data and model simplifications were made in this regard, based on the current status of the LCA literature on electromobility. Nevertheless, the case study illustrates the real value of ICLCE. The concept is expected to give a valuable contribution towards engineering future mobility systems, as it provides the wherewithal to link inputs and outputs of specific models to other models within a LCE context, i.e. with the purpose of estimating the behavior of the environmental impacts of the product system under particular design circumstances. Nevertheless, there are several limitations to the presentation of the concept that need to be addressed transparently.

Handling complex product systems such as Evs and traction batteries in LCE requires a good deal of data and models. While the prototypical implementation of the ICLCE includes a limited number of libraries of models for the manufacturing stages, product architecture and usage stage, other key stages such as the production of raw materials and the recycling stage remain largely ignored. The integration interfaces are defined, however in the case of the raw materials phase, the LCE models created were directly linked to background data from ecoinvent.

What's more, the geometrical properties of battery cells, battery packs, and vehicles are based on top-down commercial specifications of current products. Ideally the model library in ICLCE would additionally include bottom-up models representing the geometry of modern parts and components that don't necessarily exist in current designs. Within the field of LCA for electromobility, one of the largest limitations as evidenced in the current literature and from which the ICLCE concept does not escape, is the huge knowledge gap regarding the material supply chain. Along with the cell manufacturing phase, the extraction and production of battery materials is estimated to represent a significant share of the total CO_2-eq emissions of a battery system. Additionally, extraction and processing of the required materials are linked to an array of toxicity related environmental issues. Nevertheless, this estimation is based on unrealistic and, in many cases, outdated assumptions based on opaque LCI databases and LCA models. In other words, the material supply chain of traction batteries has been treated as a big black-box, with little consideration given to the complexity of these production networks. This presents two challenges. First, the modelling approaches for the assessment of the environmental impacts of raw materials for traction batteries do not capture the inherent variability of the system. For instance, new technological developments within the mining industry would alter not only the quality of the output, but essential aspects such as the electrification of energy intensive processes or transportation modes, the integration of renewable energy sources, and even technological improvements regarding the treatment of tailings and other wastes. This corresponds with the knowledge gap regarding the suitability and representativeness of impact assessment models and methods for the assessment of material supply chains for electric vehicles with regard to, for example, the spatial resolution.

Regarding the prototypical implementation of the computational system, several limitations can be discussed. First, the prototype was not conceived as a stand-alone model that can be commercialized. It doesn't provide a graphical user interface that bridges the logic with a user friendly environment. The system was entirely developed in python, and therefore requires the user to be proficient in the use of this programming language. The developed tool is demanding in terms of computational effort. The creation and estimation of 5000 product systems can take around 17 h on a regular personal laptop. Compared to the data collection process and life cycle inventory modelling in a commercial LCA software, this is still very fast. However, the development of better computational programming methods (e.g. cloud computing) will ensure faster results. Further, another important limitation of the implemented prototype is the interoperability of the tool with different background system databases. As the model was implemented using Brightway2 (Mutel 2017) for the construction of product systems and their assessment, it is (still) limited to working with ecoinvent datasets.

All in all, the most significant limitation presented in the developed prototype, and in general in LCE, is the lack of validation. Since it is not possible to know how wrong the results of a LCA are and since the implemented system relies on LCA, it is not possible to know how wrong the results of the system are. It is possible to check and compare the results against other contemporary studies addressing specific circumstances in which a battery cell is manufactured or an EV is driven, but this comparison alone won't allow us to know on what scale of factor the results are diverging from reality.

6.3 Outlook

As seen, much is still to be done if engineering activities in the field of mobility are to be equipped with the proper tools to considered sustainability issues in the development of new technologies. The following research directions can build upon the concept presented in this work.

To start, the development of a multi-model homogenization and integration platform, enabling for the retrieval of information from the disciplinary models that are useful and necessary for the LCE model. Regarding the validation issue, an ICLCE approach offers the possibility of adapting standard know-how from other research communities. There is much to learn from communities such as INCOSE, ICME and other disciplines applying Model Based Systems Engineering MBSE (Thacker et al. 2004). While a LCE model describing a real system will be difficult or impossible to calibrate, standard guidelines for the verification of coupling strategies and the validation of information flows between models can be developed. The contribution from Sargent and colleagues (Sargent 2013), to mention an example, provides a practical approach to validate and verify large scale complex simulation models. The paradigm proposed is focused on evaluating an operational, conceptual and data validation of the model. Operational validation is described by the author as the process of detecting if the output of a model "has sufficient accuracy for the model's intended purpose over the domain of the model's intended applicability". The conceptual validation refers to the processes of determining whether the theories supporting the model are able to represent the real problem reasonably. Finally, data validation refers to determining if the required data and quality of the model is available to evaluate, test and conduct experiments with the model (Sargent 2013). In LCE, these kinds of guidelines are still lacking.

The LCE research community would benefit from the development of a common unified graphic modelling language. Much can be learned from the work done by the Object Management Group (OMG) in adapting UML into SySML, a model based modelling language for the development of complex technology systems (Azevedo et al. 2009; Object Management Group 2012; Lanusse 2013). Having a common graphical modelling language in LCE would contribute towards model validation and verification through peer-reviewing, and could also foster the development of standards regarding data format exchanges and product system ontologies inside research data management clusters.

Approaches to better integrate ICLCE within the design and engineering activities of complex systems are required as well. Two possible fields are promising in this regard. First, the development of algorithms to retrospectively compute solution spaces for product systems. A solution space is defined as a multidimensional box determining permissible ranges of variations for technical parameters. Given the properties of the modelling approach presented in this work, it is realizable to generate such boxes based on environmental targets, and in this way support the robust, life cycle-oriented design and planning of a product system. Also in this direction, the combination of multidisciplinary design optimization methods with the ICLCE concept could offer important advantages towards the optimization of complex systems, while considering all possible tradeoffs and dependencies among the elements of the system.

LCE needs better charts and a better integration of statistical tools for the analysis of data and data models. The uncertainty and variability of a LCE model is not an obstacle per se, but rather an opportunity to better understand the interaction between technology and biosphere. In this regard, ICLCE approaches required more visualization techniques and tools to describe sustainability related data within a decision-making process. Some advances have been made regarding new charts (Cerdas et al. 2017a) and the use of visual analytics and visualization devices (Juraschek et al. 2018; Kaluza et al. 2018).

The integration of the ICLCE concept into sources providing real or near time data into the ICLCE concept is a promising approach for increasing the resolution of the data required in LCE activities, and also to enhance the interaction with the results during the execution of visual and data analytics processes. Some first concepts are proposed in (Cerdas et al. 2017b; Hagen et al. 2020). ICLCE offers the opportunity to dynamically develop life cycle inventories that can be integrated with in-house data collection approaches as in (Koffler et al. 2008). As the block-chain technology increasingly gains relevance in the field of sustainability and transparency, several publications have recently started to prepare the field for a block-chain LCA (Teh et al. 2020; Venkatesh et al. 2020; Zhang et al. 2020). ICLCE intrinsic data management and traceability properties offer, in combination with block chain technologies, good potentials to support the establishment of multi-stakeholder secured life cycle inventory development, while reducing issues of privacy and confidentiality during the data exchange process between different companies.

This can also be followed for the development of model-based background data sets that can be linked to ICLCE models in the future, without the need of developing background systems models as generically described in (Hick et al. 2019). The vision is that a provider of background system models could be adapted and linked in the ICLCE model for a specific technology, similarly to how a datasets provider such as ecoinvent operates. Regarding the usability of the concept, several computational developments can be pursued in order to enhance its applicability. In this regard, a further investigation regarding the potential business models throughout the data and mode pipeline in ICLCE is necessary to ensure the continuous growth and updatability of the system.

The field of mobility and LCE is in urgent need of better models and data. This is key for the success of any system analysis tool that aims to give decision making support regarding sustainability aspects. This implies that there is a need for models in each of the life cycle stages of a traction battery, models for future battery materials, and models for the study of the usage and recycling stages. Finally, the modelling approach can be transferred to many other application fields. Particularly relevant, regarding the current developed model library, are systems in the field of future mobility, such as aviation and other energy production, and storage systems such as fuel cell systems and hydrogen.

References

Azevedo K, Bras B, Doshi S et al (2009) Modeling sustainability of complex systems: a multi-scale framework using SysML. ASME 2009

Cerdas F, Kaluza A, Erkisi-Arici S et al (2017a) Improved visualization in LCA through the application of cluster heat maps. Procedia CIRP 00:732–737. https://doi.org/10.1016/j.procir.2016.11.160

Cerdas F, Thiede S, Juraschek M et al (2017b) Shop-floor life cycle assessment. In: Procedia CIRP, pp 393–398

Hagen J, Büth L, Haupt J et al (2020) Live LCA in learning factories: real time assessment of product life cycles environmental impacts. Procedia Manuf 45:128–133. https://doi.org/10.1016/j.promfg.2020.04.083

Hauschild MZ, Herrmann C, Kara S (2017) An integrated framework for life cycle engineering. Procedia CIRP 61:2–9. https://doi.org/10.1016/j.procir.2016.11.257

Hick H, Bajzek M, Faustmann C (2019) Definition of a system model for model-based development. SN Appl Sci 1:1–15. https://doi.org/10.1007/s42452-019-1069-0

Horstemeyer MF, Sahay S (2018) Definition of ICME. In: Horstemeyer MF (ed) Integrated computational materials engineering (ICME) for metals. Wiley, Hoboken, NJ, USA, pp 1–17

Juraschek M, Büth L, Cerdas F et al (2018) Exploring the potentials of mixed reality for life cycle engineering. In: Procedia CIRP

Kaluza A, Gellrich S, Cerdas F et al (2018) Life cycle engineering based on visual analytics. In: Procedia CIRP

Koffler C, Krinke S, Schebek L, Buchgeister J (2008) Volkswagen slimLCI: a procedure for streamlined inventory modelling within life cycle assessment of vehicles. Int J Veh Des 46:172. https://doi.org/10.1504/IJVD.2008.017181

Lanusse A (2013) Introduction to sysml

Mutel C (2017) Brightway : an open source framework for life cycle assessment. 47:11–12. https://doi.org/10.21105/joss.00236

Object Management Group (2012) OMG system modeling language (OMG SysML™). 272

Sargent RG (2013) Verification and validation of simulation models. J Simul 7:12–24. https://doi.org/10.1057/jos.2012.20

Teh D, Khan T, Corbitt B, Ong CE (2020) Sustainability strategy and blockchain-enabled life cycle assessment: a focus on materials industry. Environ Syst Decis. https://doi.org/10.1007/s10669-020-09761-4

Thacker BH, Doebling SW, Hemez FM et al (2004) Concepts of model verification and validation. Concepts Model Verif Valid 41. https://doi.org/10.2172/835920

Venkatesh VG, Kang K, Wang B et al (2020) System architecture for blockchain based transparency of supply chain social sustainability. Robot Comput Integr Manuf 63. https://doi.org/10.1016/j.rcim.2019.101896

Zhang A, Zhong RY, Farooque M et al (2020) Blockchain-based life cycle assessment: An implementation framework and system architecture. Resour Conserv Recycl 152. https://doi.org/10.1016/j.resconrec.2019.104512

Appendix

A. Representation of the structure of a product object in ICLCE

F. Cerdas, *Integrated Computational Life Cycle Engineering for Traction Batteries*,
Sustainable Production, Life Cycle Engineering and Management,
https://doi.org/10.1007/978-3-030-82934-6

- **product_object**
    - ✓ product_object_uuid
    - ✓ database: foreground_system/product [vehicle, pack or cell]
    - + **product_performance**
    - – **product_architecture**
        - – **raw_materials**
            - ✓ raw_materials_uuid
            - ✓ database: foreground_system/raw_materials
            - – **material_supply_chain_1**
                - ✓ material_supply_chain_uuid
                - ✓ database: .../raw_materials/material1
                - + **processes**
                - + **transportation**
                - + **...**
            - + **material_supply_chain_2**
            - + **material_supply_chain_3**
            - + **material_supply_chain_4**
            - + **material_supply_chain_5**
            - + **material_supply_chain_6**
            - + **...**
            - + **material_supply_chain_n**
        - – **manufacturing**
            - ✓ manufacturing_uuid
            - ✓ database: foreground/manufacturing
            - + **factory**
            - – **process_chain**
                - ✓ process_chain_uuid
                - ✓ database: foreground/process_chain
                - – **process_1**
                    - ✓ process_uuid
                    - ✓ database: foreground/manufacturing/processes
                - + **...**
                - + **process_x**
        - – **Usage**
            - ✓ usage_uuid
            - ✓ database:foreground/usage
            - + **energy_consumers**
            - + **life_time_degraders**
        - – **Recycling**
            - ✓ recycling_uuid
            - ✓ database: foreground/recycling
            - + **factory**
            - – **process_chain**
                - – **process_1**
                    - ✓ process_uuid
                    - ✓ database: foreground/recycling/processes
                - + **...**
                - + **process_x**

B. Model coupling matrix for the implemented models in the ICLE. Whole system.

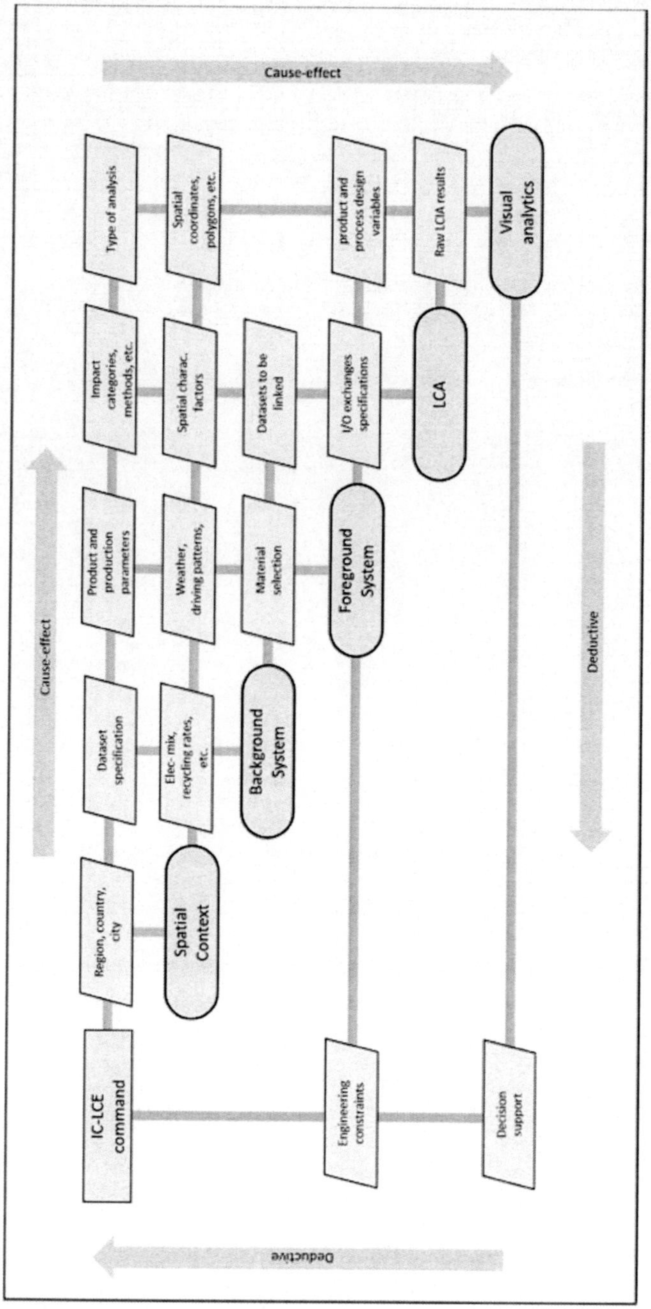
Cause-effect
Cause-effect
Deductive
Deductive
IC-LCE command
Region, country, city
Dataset specification
Product and production parameters
Impact categories, methods, etc.
Type of analysis
Spatial Context
Elec- mix, recycling rates, etc.
Weather, driving patterns,
Spatial charac. factors
Spatial coordinates, polygons, etc.
Background System
Material selection
Datasets to be linked
Foreground System
I/O exchanges specifications
product and process design variables
LCA
Raw LCIA results
Visual analytics
Engineering constraints
Decision support

C. Model coupling matrix for the implemented models in the ICLE. Foreground system.

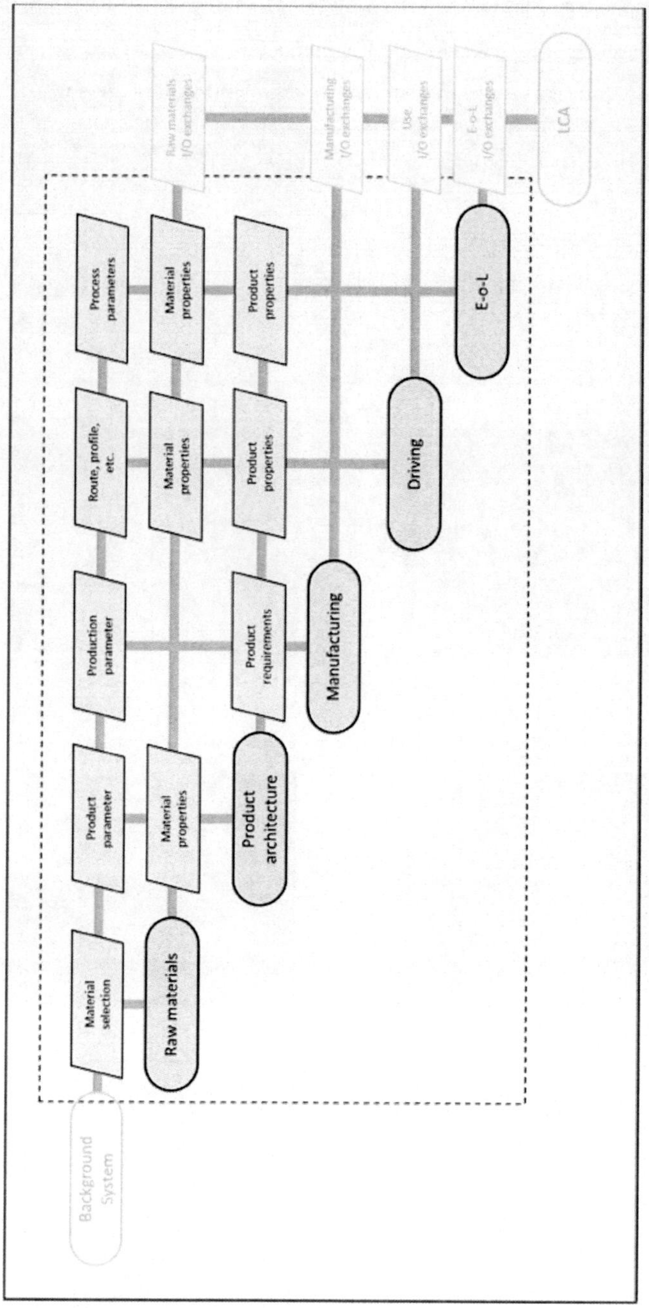
Background System
Material selection
Raw materials
Product parameter
Material properties
Product architecture
Production parameter
Product requirements
Manufacturing
Route, profile, etc.
Material properties
Product properties
Driving
Process parameters
Material properties
Product properties
E-o-L
Raw materials I/O exchanges
Manufacturing I/O exchanges
Use I/O exchanges
E-o-L I/O exchanges
LCA

D. Estimated temperature profiles for all the cities considered in the usage stage.

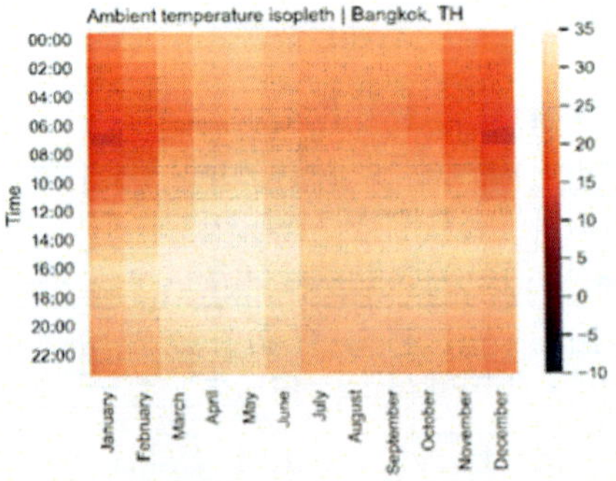

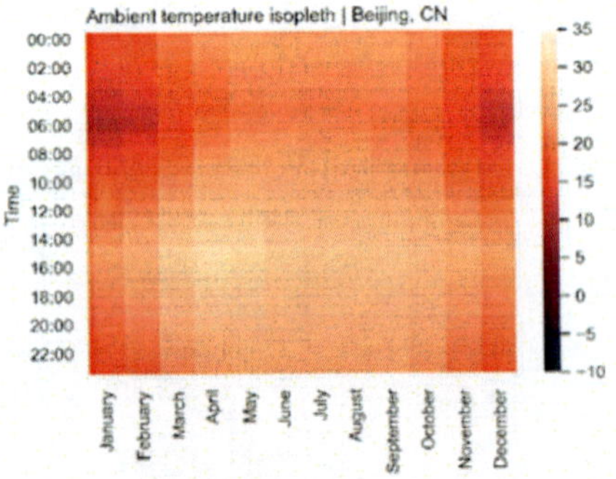

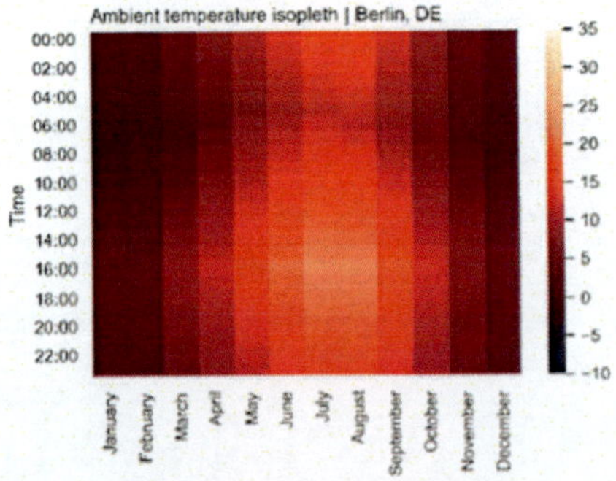

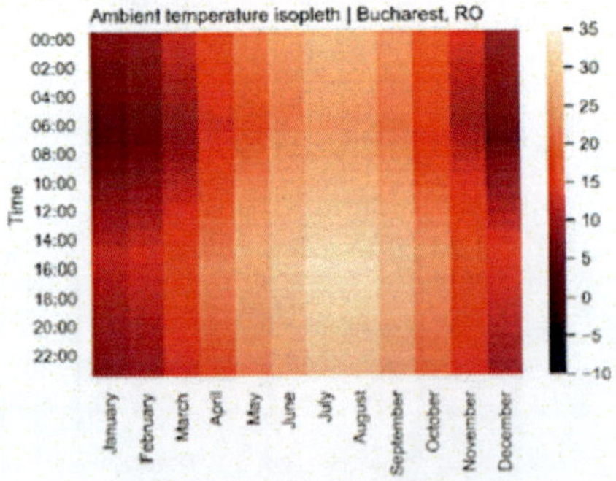

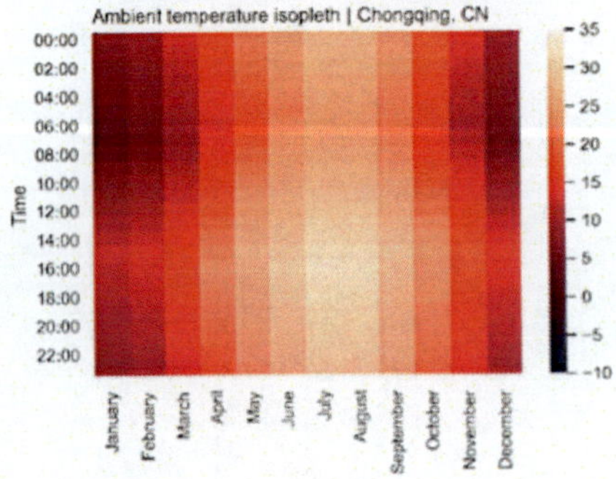

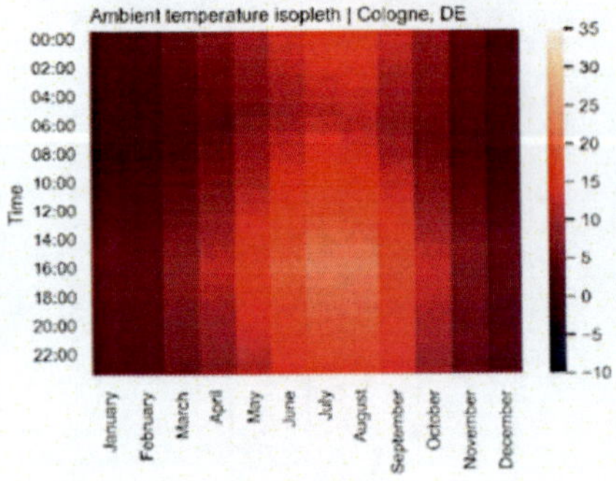

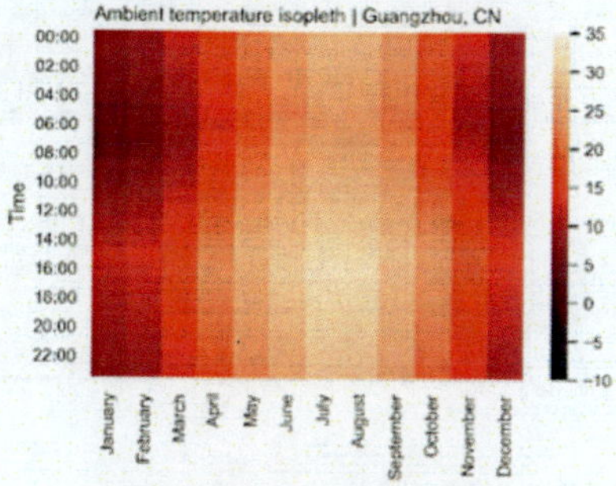
Ambient temperature isopleth | Guangzhou, CN
Time
00:00 02:00 04:00 06:00 08:00 10:00 12:00 14:00 16:00 18:00 20:00 22:00
January February March April May June July August September October November December
35 30 25 20 15 10 5 0 -5 -10

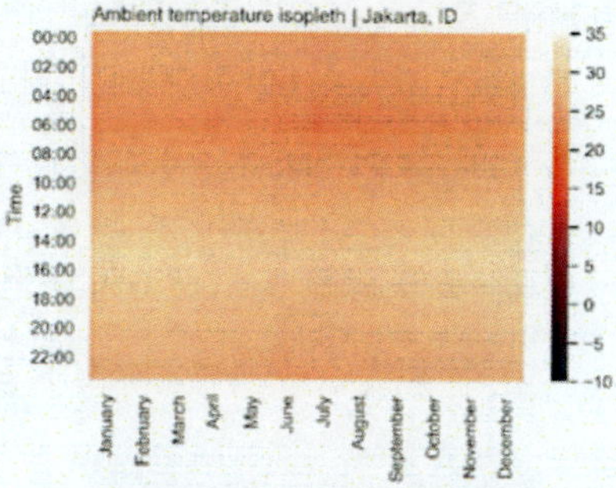
Ambient temperature isopleth | Jakarta, ID
Time
00:00 02:00 04:00 06:00 08:00 10:00 12:00 14:00 16:00 18:00 20:00 22:00
January February March April May June July August September October November December
35 30 25 20 15 10 5 0 -5 -10

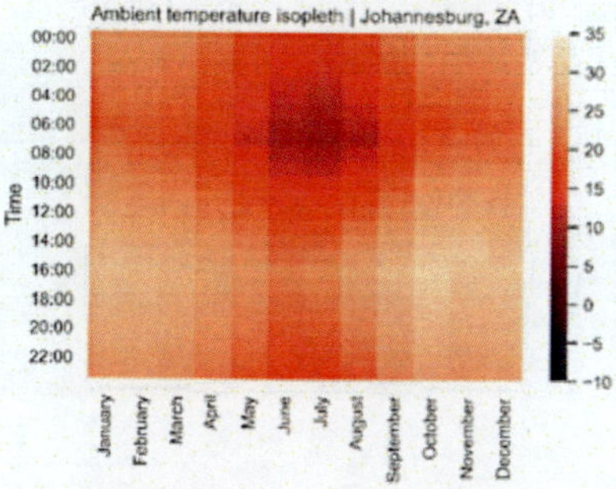
Ambient temperature isopleth | Johannesburg, ZA
Time
00:00 02:00 04:00 06:00 08:00 10:00 12:00 14:00 16:00 18:00 20:00 22:00
January February March April May June July August September October November December
35 30 25 20 15 10 5 0 -5 -10

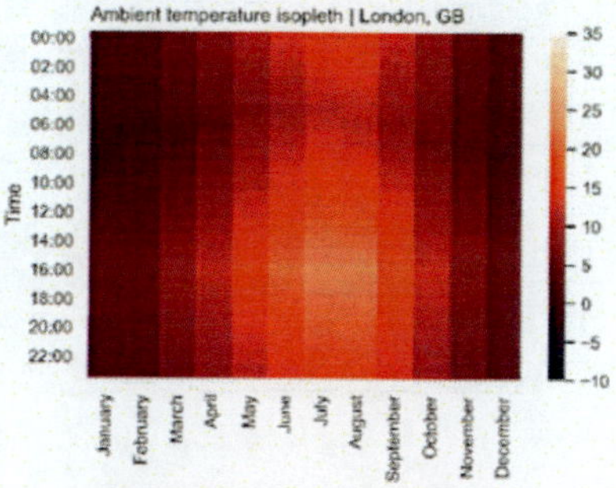
Ambient temperature isopleth | London, GB
Time
00:00 02:00 04:00 06:00 08:00 10:00 12:00 14:00 16:00 18:00 20:00 22:00
January February March April May June July August September October November December
35 30 25 20 15 10 5 0 -5 -10

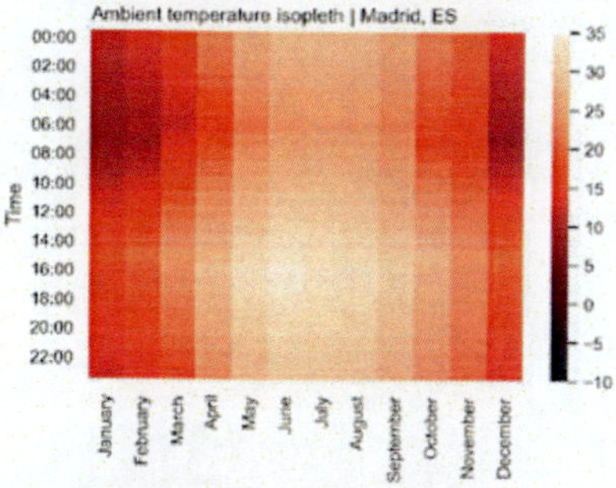
Ambient temperature isopleth | Madrid, ES
Time
00:00 02:00 04:00 06:00 08:00 10:00 12:00 14:00 16:00 18:00 20:00 22:00
January February March April May June July August September October November December
35 30 25 20 15 10 5 0 -5 -10

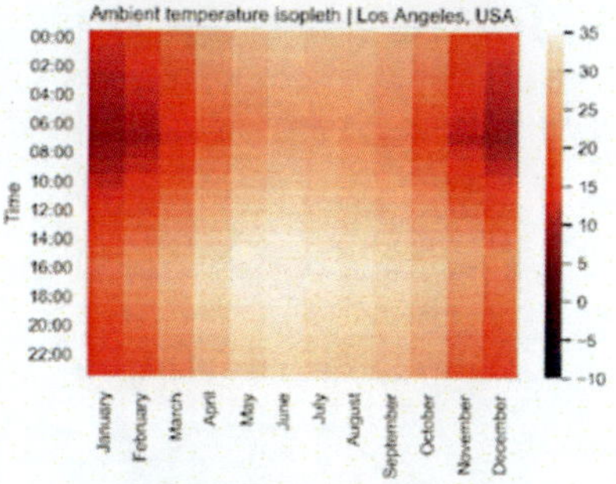
Ambient temperature isopleth | Los Angeles, USA
Time
00:00 02:00 04:00 06:00 08:00 10:00 12:00 14:00 16:00 18:00 20:00 22:00
January February March April May June July August September October November December
35 30 25 20 15 10 5 0 -5 -10

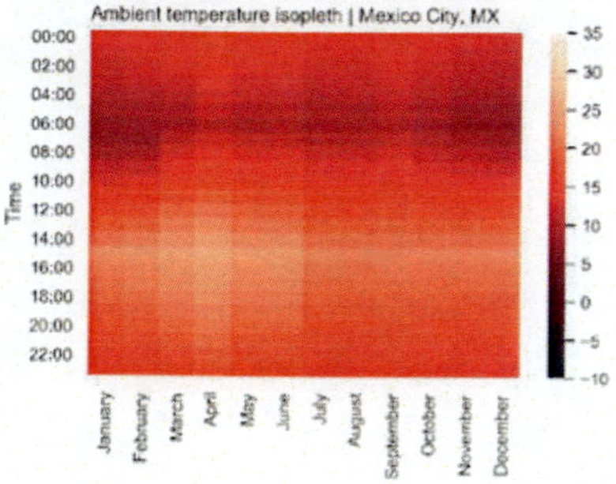
Ambient temperature isopleth | Mexico City, MX
Time
00:00 02:00 04:00 06:00 08:00 10:00 12:00 14:00 16:00 18:00 20:00 22:00
January February March April May June July August September October November December
35 30 25 20 15 10 5 0 -5 -10

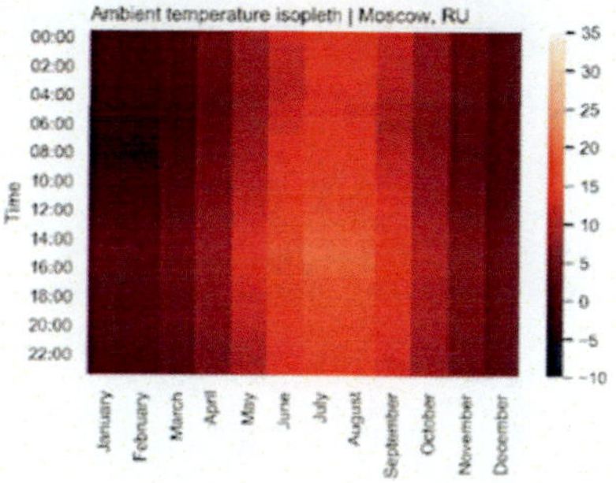
Ambient temperature isopleth | Moscow, RU
Time
00:00 02:00 04:00 06:00 08:00 10:00 12:00 14:00 16:00 18:00 20:00 22:00
January February March April May June July August September October November December
35 30 25 20 15 10 5 0 -5 -10

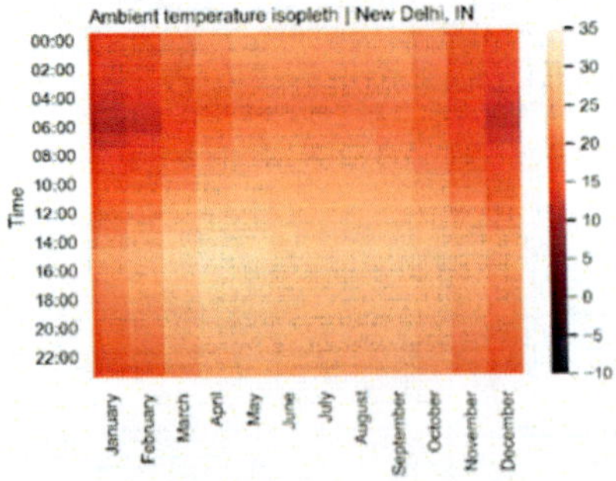
Ambient temperature isopleth | New Delhi, IN
Time
00:00 02:00 04:00 06:00 08:00 10:00 12:00 14:00 16:00 18:00 20:00 22:00
January February March April May June July August September October November December
35 30 25 20 15 10 5 0 −5 −10

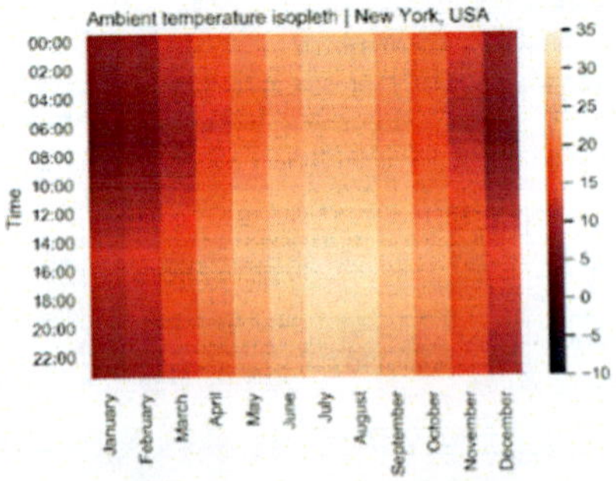
Ambient temperature isopleth | New York, USA
Time
00:00 02:00 04:00 06:00 08:00 10:00 12:00 14:00 16:00 18:00 20:00 22:00
January February March April May June July August September October November December
35 30 25 20 15 10 5 0 −5 −10

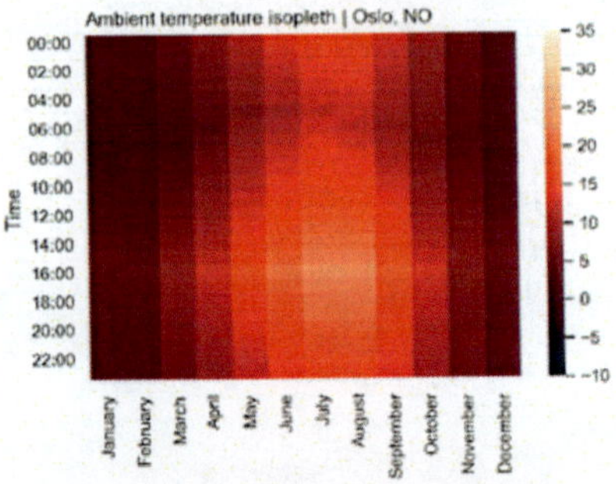
Ambient temperature isopleth | Oslo, NO
Time
00:00 02:00 04:00 06:00 08:00 10:00 12:00 14:00 16:00 18:00 20:00 22:00
January February March April May June July August September October November December
35 30 25 20 15 10 5 0 −5 −10

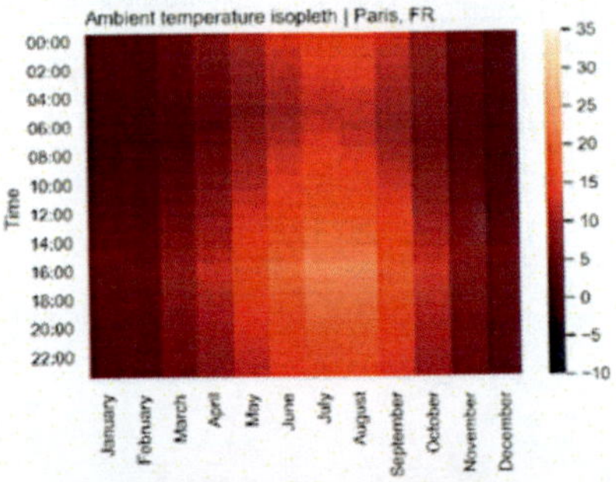
Ambient temperature isopleth | Paris, FR
Time
00:00 02:00 04:00 06:00 08:00 10:00 12:00 14:00 16:00 18:00 20:00 22:00
January February March April May June July August September October November December
35 30 25 20 15 10 5 0 −5 −10

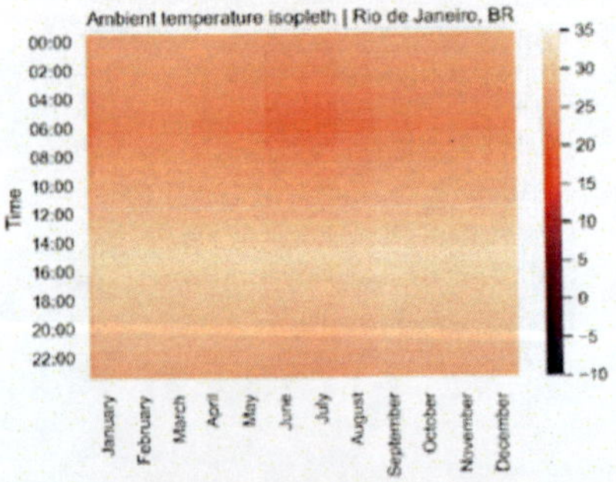
Ambient temperature isopleth | Rio de Janeiro, BR
Time
00:00 02:00 04:00 06:00 08:00 10:00 12:00 14:00 16:00 18:00 20:00 22:00
January February March April May June July August September October November December
35 30 25 20 15 10 5 0 −5 −10

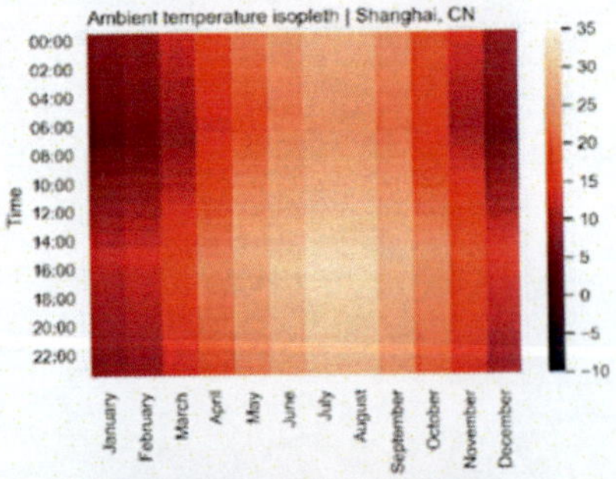
Ambient temperature isopleth | Shanghai, CN
Time
00:00 02:00 04:00 06:00 08:00 10:00 12:00 14:00 16:00 18:00 20:00 22:00
January February March April May June July August September October November December
35 30 25 20 15 10 5 0 −5 −10

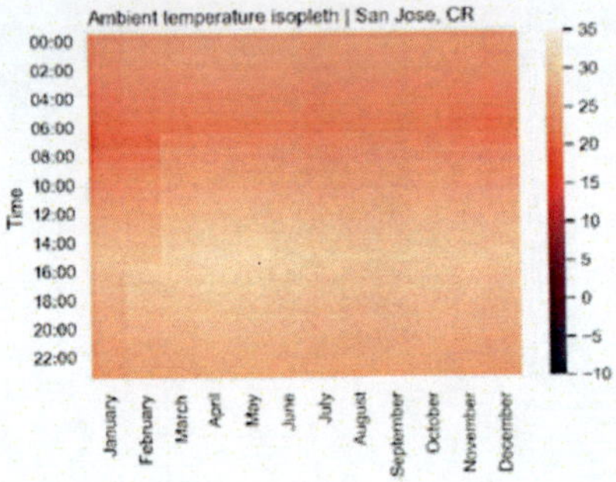
Ambient temperature isopleth | San Jose, CR
Time
00:00 02:00 04:00 06:00 08:00 10:00 12:00 14:00 16:00 18:00 20:00 22:00
January February March April May June July August September October November December
35 30 25 20 15 10 5 0 −5 −10

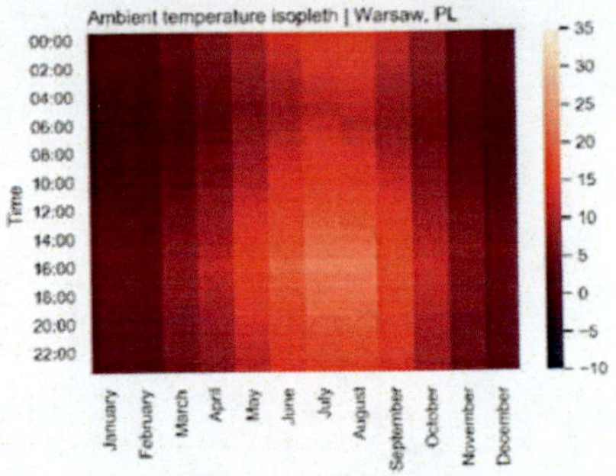
Ambient temperature isopleth | Warsaw, PL
Time
00:00 02:00 04:00 06:00 08:00 10:00 12:00 14:00 16:00 18:00 20:00 22:00
January February March April May June July August September October November December
35 30 25 20 15 10 5 0 −5 −10

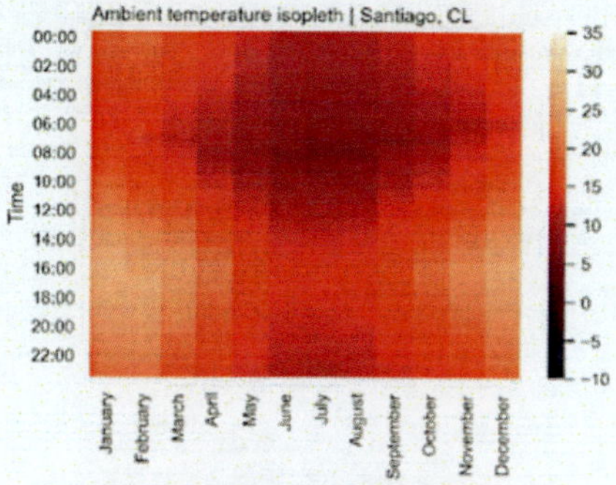

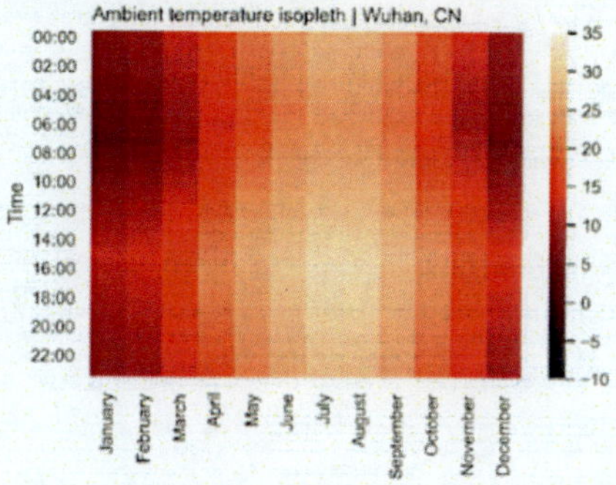

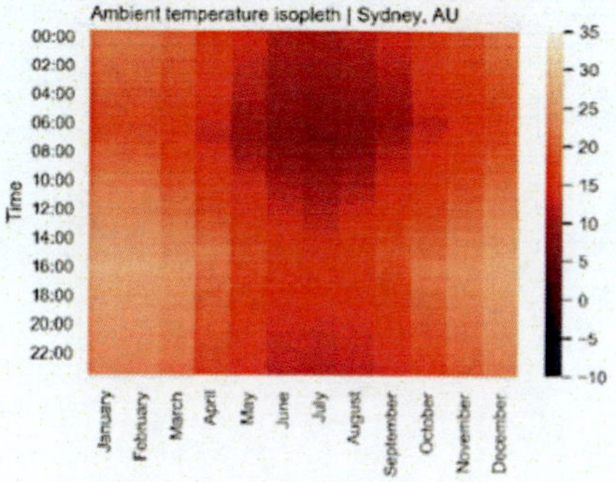

E. Speed profiles calculated for all the cities considered in usage stage.

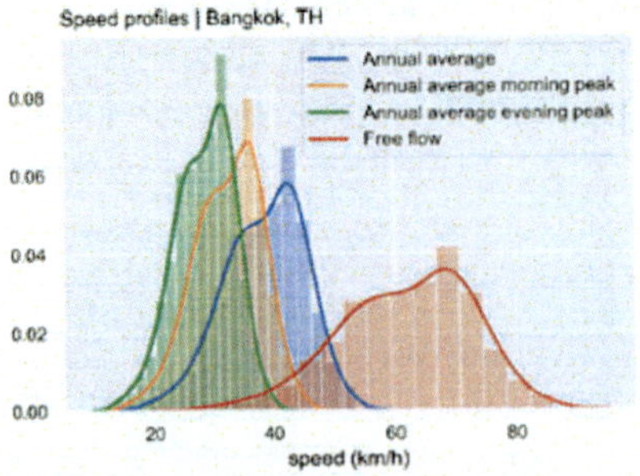
Speed profiles | Bangkok, TH
Annual average
Annual average morning peak
Annual average evening peak
Free flow
speed (km/h)

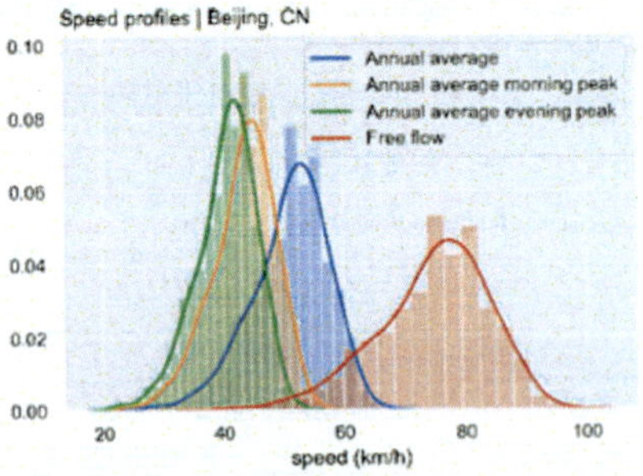
Speed profiles | Beijing, CN
Annual average
Annual average morning peak
Annual average evening peak
Free flow
speed (km/h)

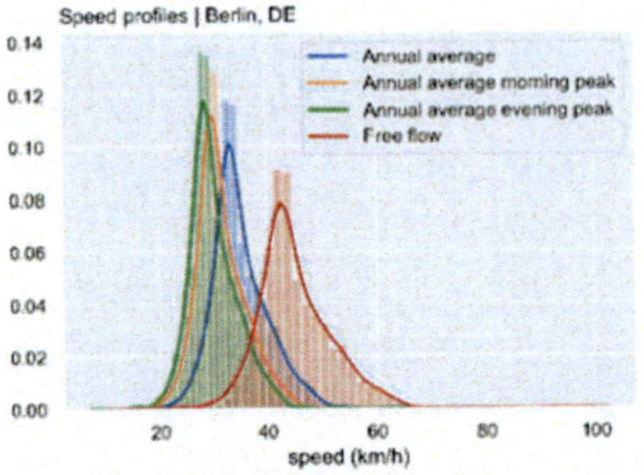
Speed profiles | Berlin, DE
Annual average
Annual average morning peak
Annual average evening peak
Free flow
speed (km/h)

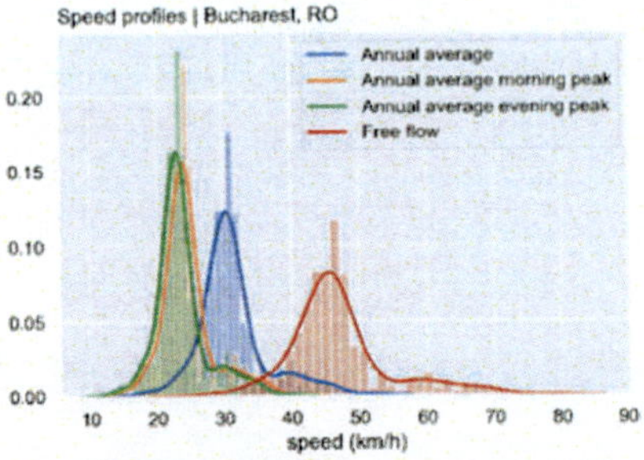
Speed profiles | Bucharest, RO
Annual average
Annual average morning peak
Annual average evening peak
Free flow
speed (km/h)

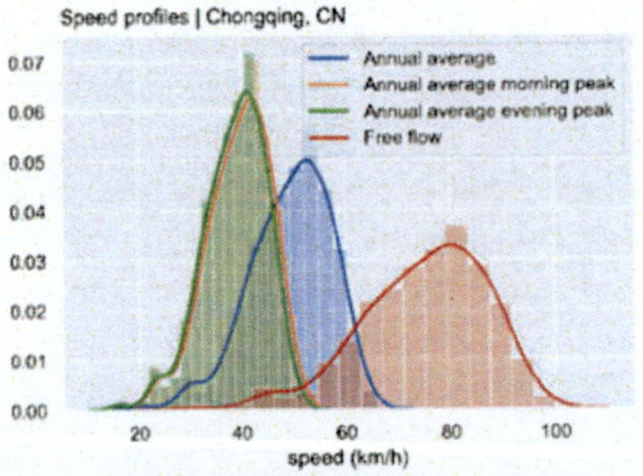
Speed profiles | Chongqing, CN
Annual average
Annual average morning peak
Annual average evening peak
Free flow
speed (km/h)

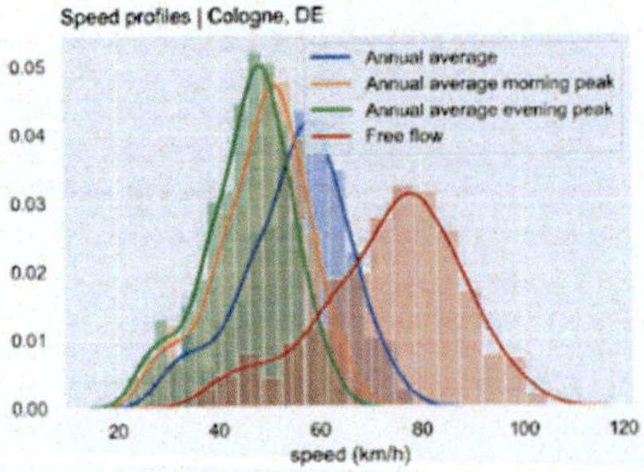
Speed profiles | Cologne, DE
Annual average
Annual average morning peak
Annual average evening peak
Free flow
speed (km/h)

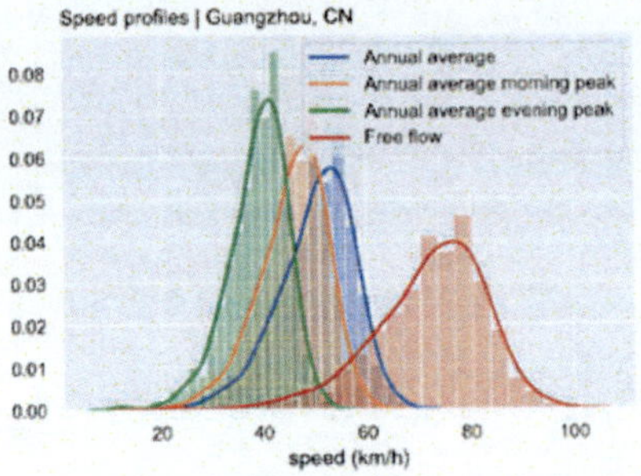
Speed profiles | Guangzhou, CN
Annual average
Annual average morning peak
Annual average evening peak
Free flow
speed (km/h)

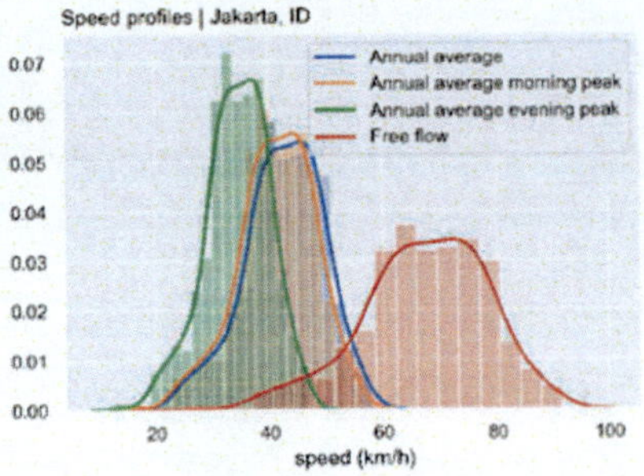
Speed profiles | Jakarta, ID
Annual average
Annual average morning peak
Annual average evening peak
Free flow
speed (km/h)

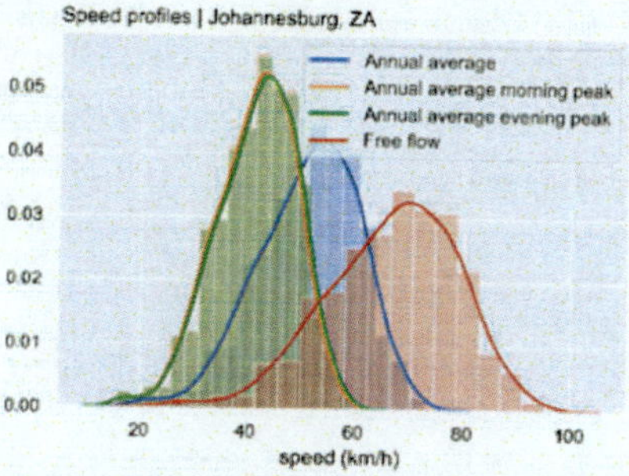
Speed profiles | Johannesburg, ZA
Annual average
Annual average morning peak
Annual average evening peak
Free flow
speed (km/h)

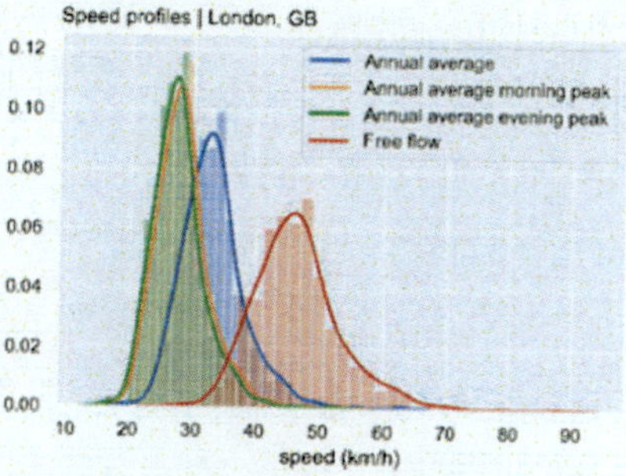
Speed profiles | London, GB
Annual average
Annual average morning peak
Annual average evening peak
Free flow
speed (km/h)

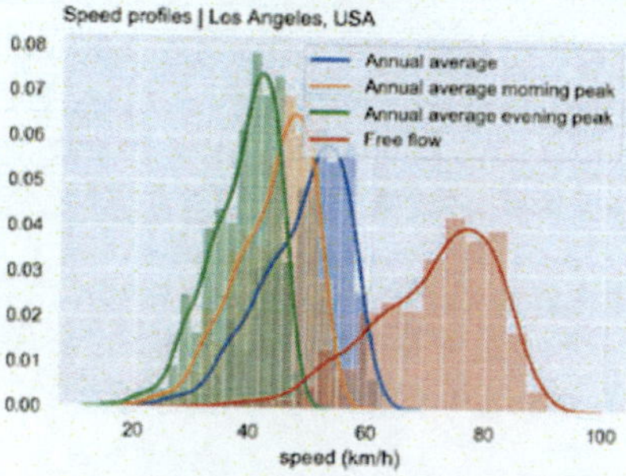
Speed profiles | Los Angeles, USA
Annual average
Annual average morning peak
Annual average evening peak
Free flow
speed (km/h)

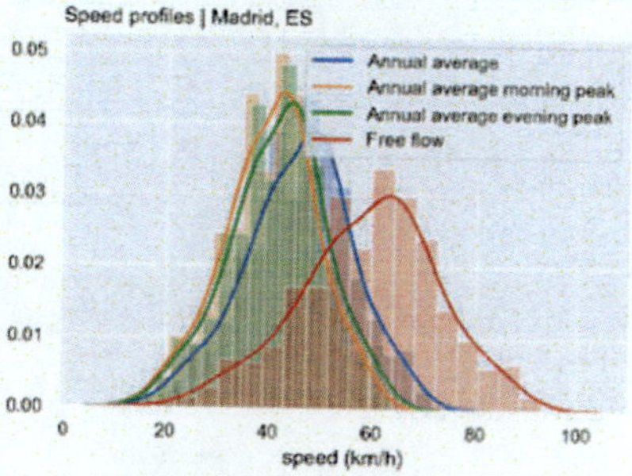
Speed profiles | Madrid, ES
Annual average
Annual average morning peak
Annual average evening peak
Free flow
speed (km/h)

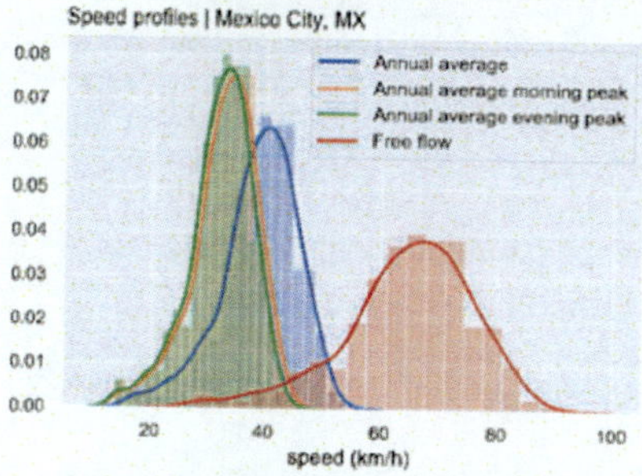
Speed profiles | Mexico City, MX
Annual average
Annual average morning peak
Annual average evening peak
Free flow
speed (km/h)

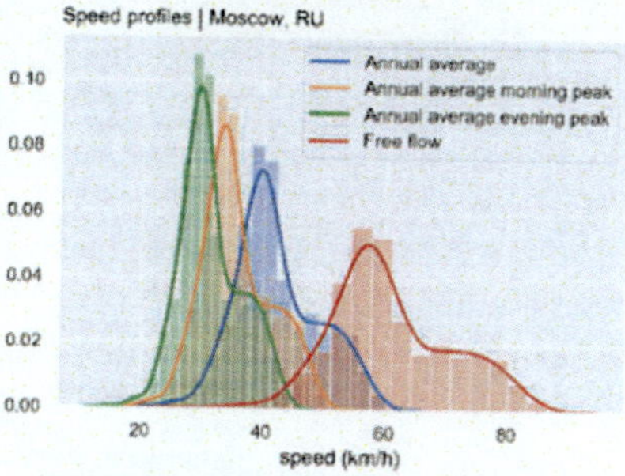
Speed profiles | Moscow, RU
Annual average
Annual average morning peak
Annual average evening peak
Free flow
speed (km/h)

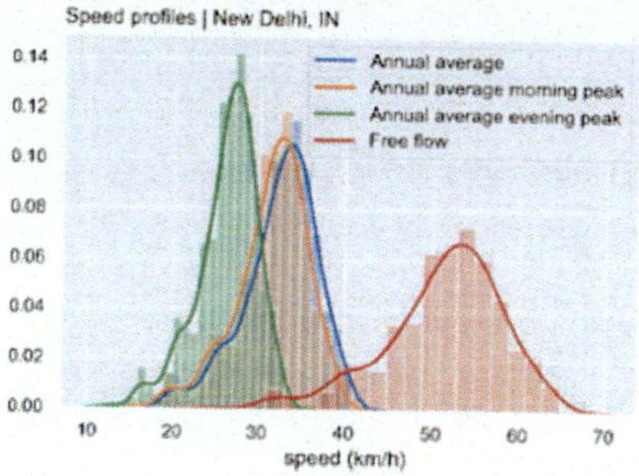
Speed profiles | New Delhi, IN
Annual average
Annual average morning peak
Annual average evening peak
Free flow
speed (km/h)

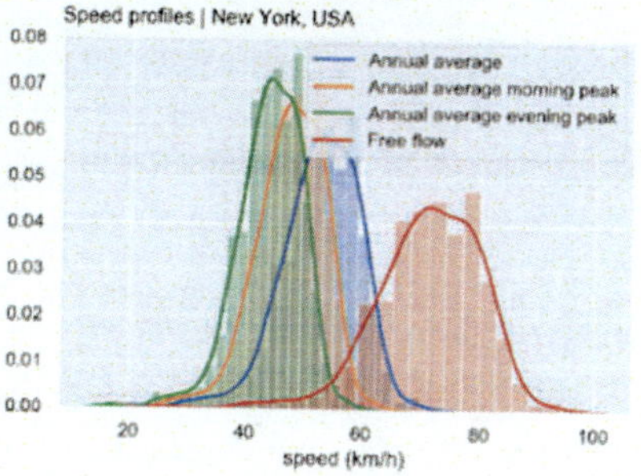
Speed profiles | New York, USA
Annual average
Annual average morning peak
Annual average evening peak
Free flow
speed (km/h)

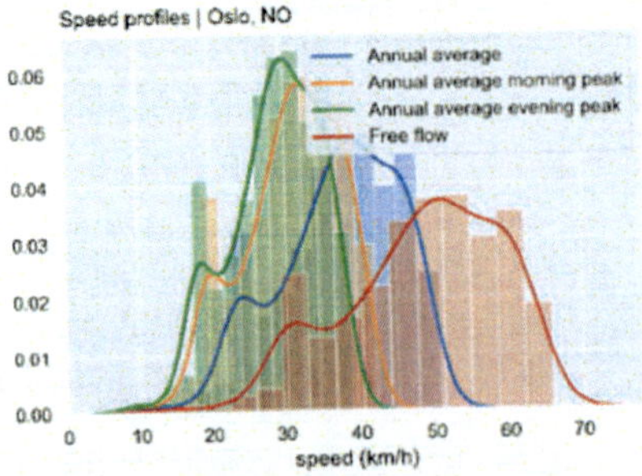
Speed profiles | Oslo, NO
Annual average
Annual average morning peak
Annual average evening peak
Free flow
speed (km/h)

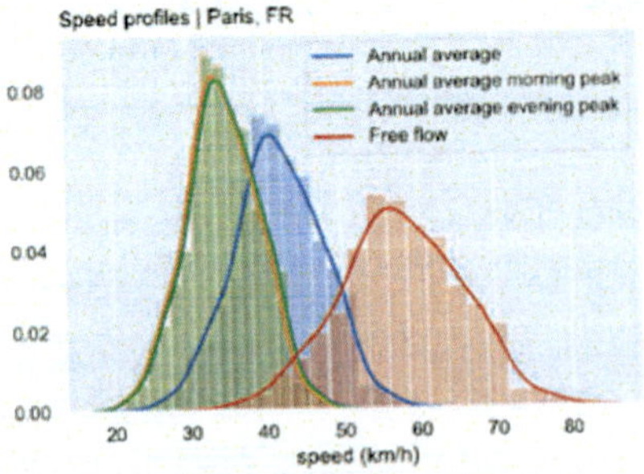
Speed profiles | Paris, FR
Annual average
Annual average morning peak
Annual average evening peak
Free flow
speed (km/h)

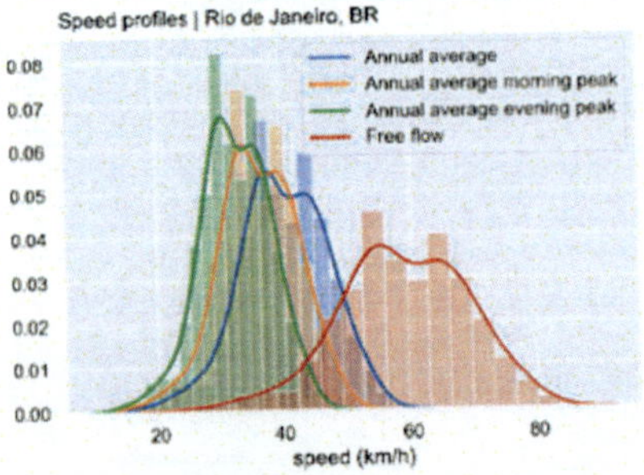
Speed profiles | Rio de Janeiro, BR
Annual average
Annual average morning peak
Annual average evening peak
Free flow
speed (km/h)

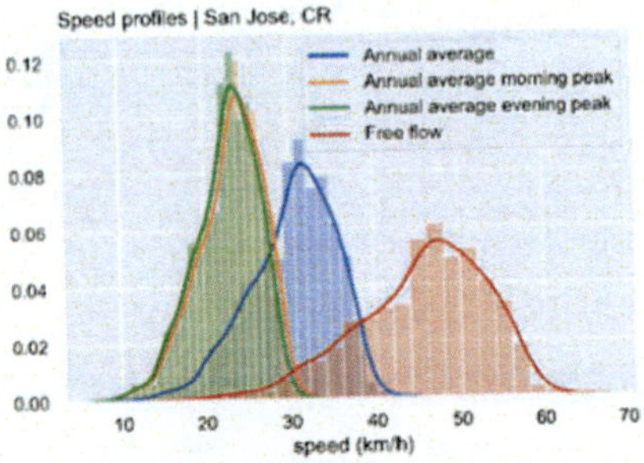
Speed profiles | San Jose, CR
Annual average
Annual average morning peak
Annual average evening peak
Free flow
speed (km/h)

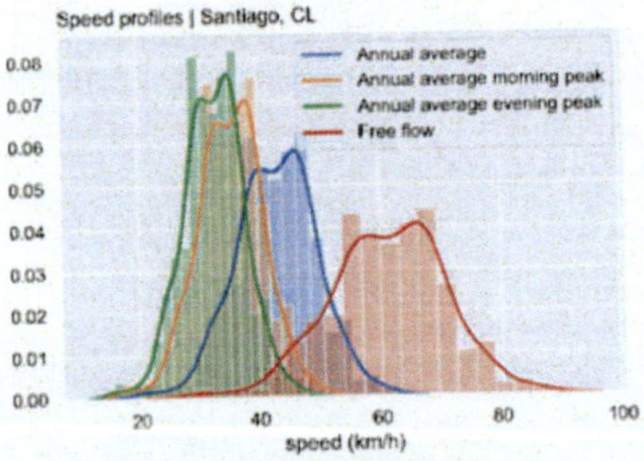
Speed profiles | Santiago, CL
Annual average
Annual average morning peak
Annual average evening peak
Free flow
speed (km/h)

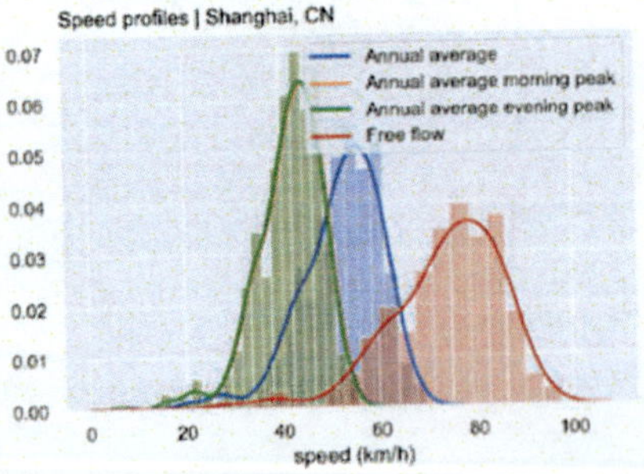
Speed profiles | Shanghai, CN
Annual average
Annual average morning peak
Annual average evening peak
Free flow
speed (km/h)

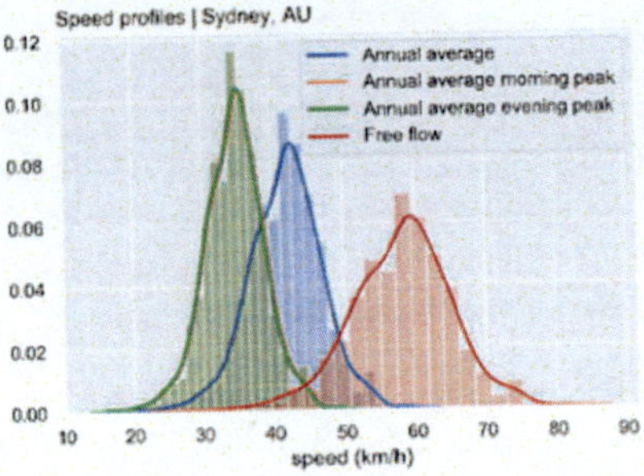
Speed profiles | Sydney, AU
Annual average
Annual average morning peak
Annual average evening peak
Free flow
speed (km/h)

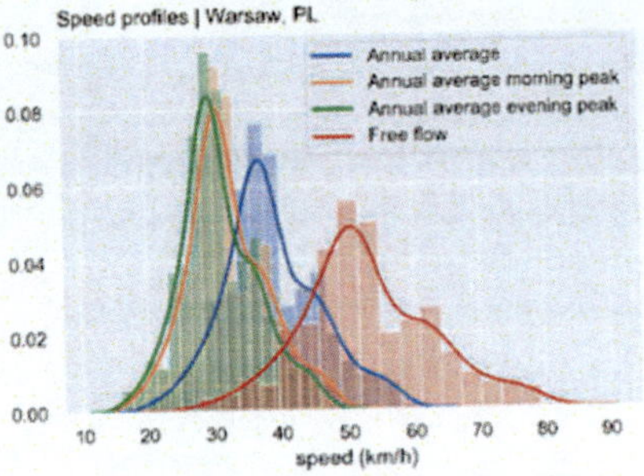
Speed profiles | Warsaw, PL
Annual average
Annual average morning peak
Annual average evening peak
Free flow
speed (km/h)

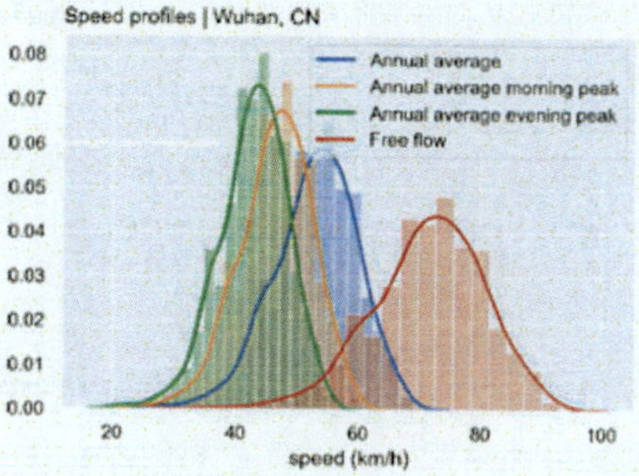

F. Energy demand profile mix for all the cities considered in the usage stage. Vehicle of the B segment with a NMC-622 battery pack.

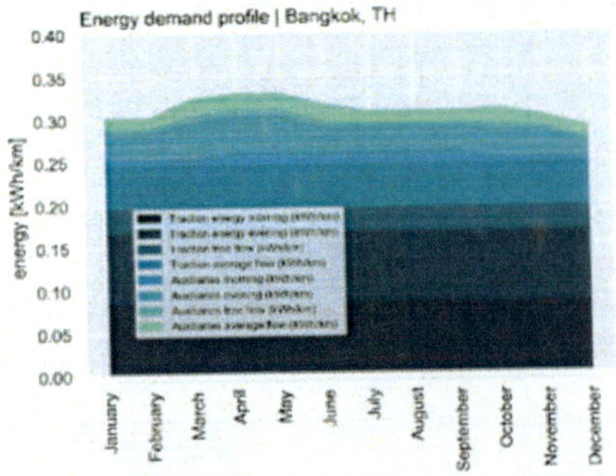
Energy demand profile | Bangkok, TH
energy [kWh/km]
0.00 0.05 0.10 0.15 0.20 0.25 0.30 0.35 0.40
January February March April May June July August September October November December

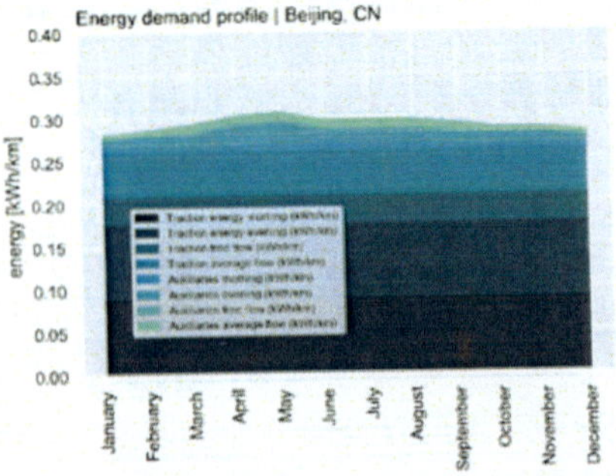
Energy demand profile | Beijing, CN
energy [kWh/km]
0.00 0.05 0.10 0.15 0.20 0.25 0.30 0.35 0.40
January February March April May June July August September October November December

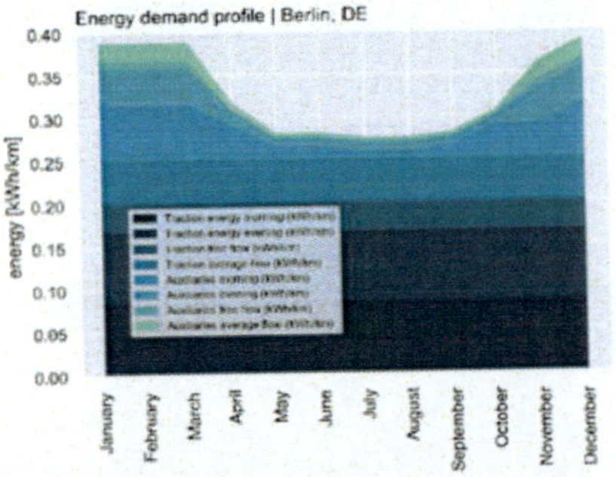
Energy demand profile | Berlin, DE
energy [kWh/km]
0.00 0.05 0.10 0.15 0.20 0.25 0.30 0.35 0.40
January February March April May June July August September October November December

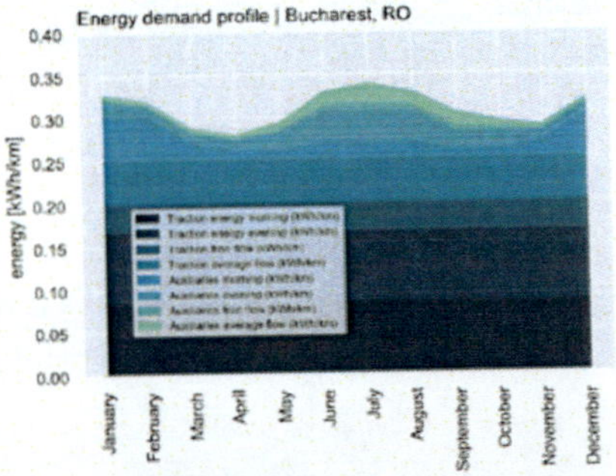
Energy demand profile | Bucharest, RO
energy [kWh/km]
0.00 0.05 0.10 0.15 0.20 0.25 0.30 0.35 0.40
January February March April May June July August September October November December

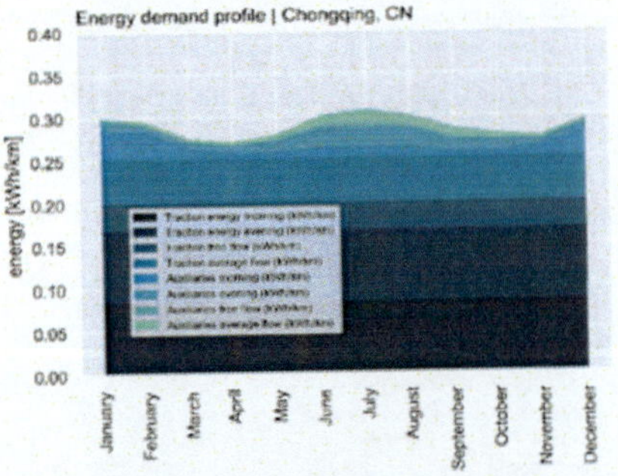
Energy demand profile | Chongqing, CN
energy [kWh/km]
0.00 0.05 0.10 0.15 0.20 0.25 0.30 0.35 0.40
January February March April May June July August September October November December

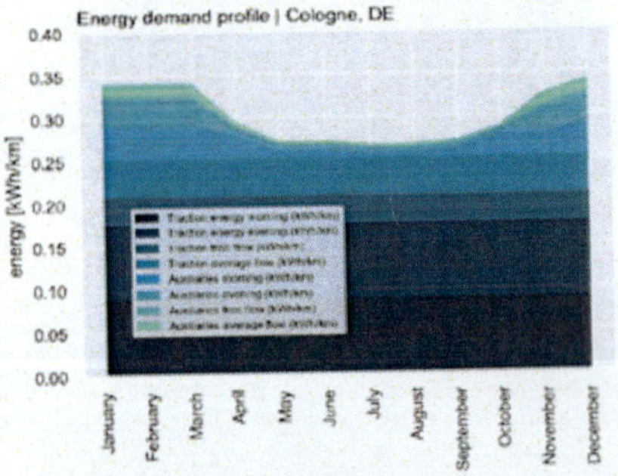
Energy demand profile | Cologne, DE
energy [kWh/km]
0.00 0.05 0.10 0.15 0.20 0.25 0.30 0.35 0.40
January February March April May June July August September October November December

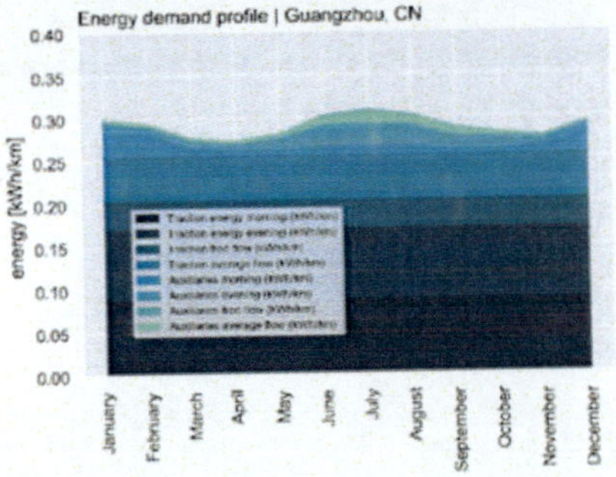
Energy demand profile | Guangzhou, CN
energy [kWh/km]
0.00 0.05 0.10 0.15 0.20 0.25 0.30 0.35 0.40
January February March April May June July August September October November December

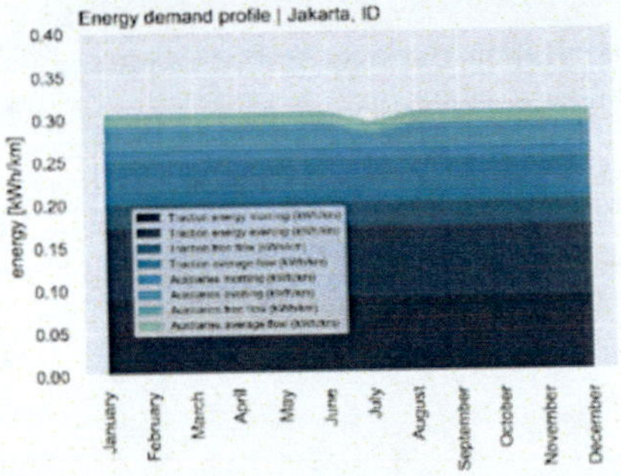
Energy demand profile | Jakarta, ID
energy [kWh/km]
0.00 0.05 0.10 0.15 0.20 0.25 0.30 0.35 0.40
January February March April May June July August September October November December

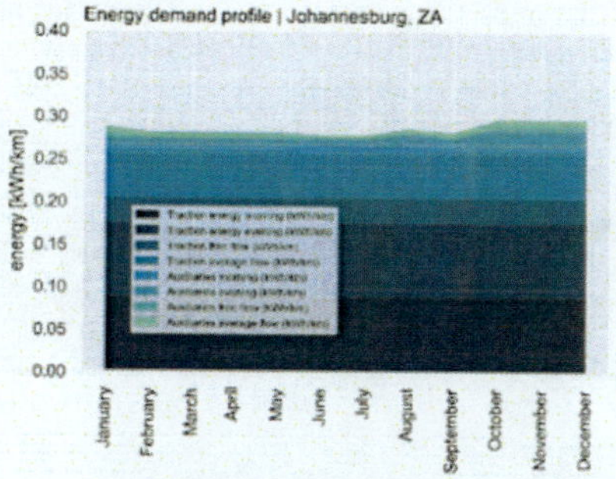
Energy demand profile | Johannesburg, ZA
energy [kWh/km]
0.40
0.35
0.30
0.25
0.20
0.15
0.10
0.05
0.00
January
February
March
April
May
June
July
August
September
October
November
December

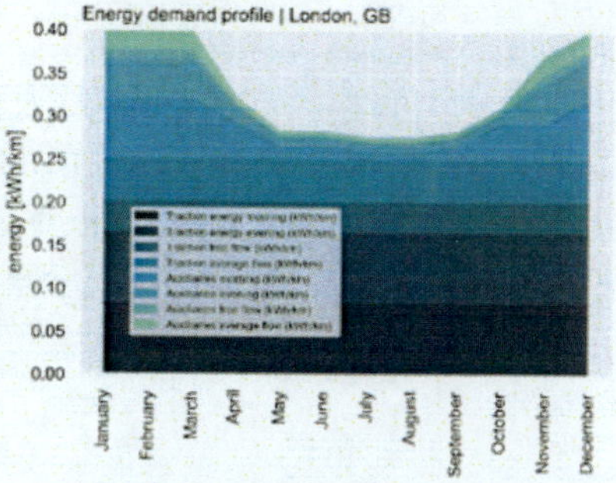
Energy demand profile | London, GB
energy [kWh/km]
0.40
0.35
0.30
0.25
0.20
0.15
0.10
0.05
0.00
January
February
March
April
May
June
July
August
September
October
November
December

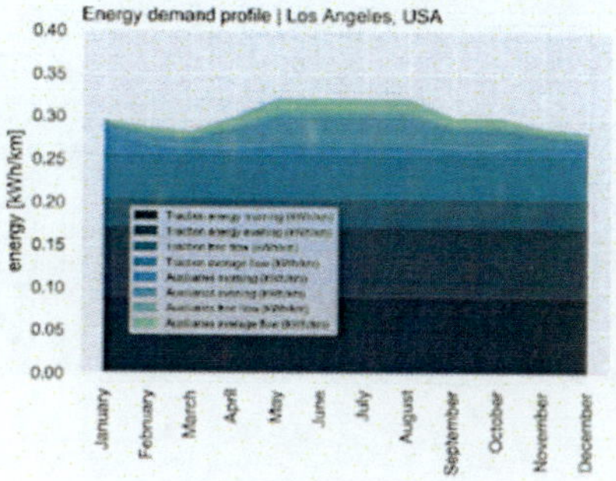
Energy demand profile | Los Angeles, USA
energy [kWh/km]
0.40
0.35
0.30
0.25
0.20
0.15
0.10
0.05
0.00
January
February
March
April
May
June
July
August
September
October
November
December

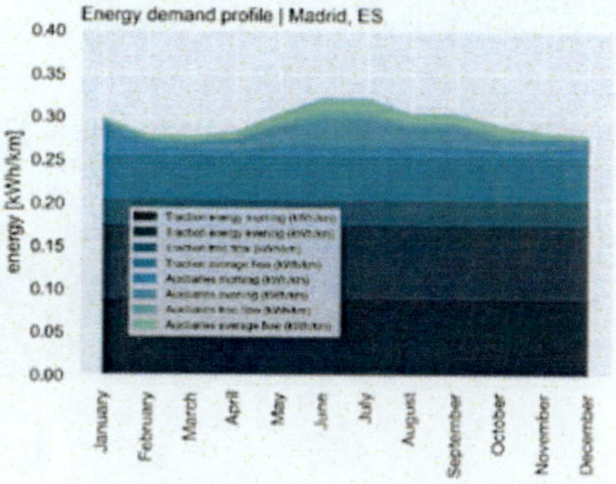
Energy demand profile | Madrid, ES
energy [kWh/km]
0.40
0.35
0.30
0.25
0.20
0.15
0.10
0.05
0.00
January
February
March
April
May
June
July
August
September
October
November
December

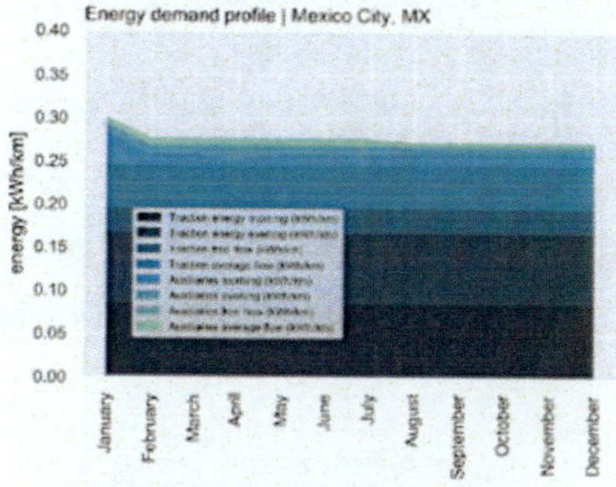
Energy demand profile | Mexico City, MX
energy [kWh/km]
0.40
0.35
0.30
0.25
0.20
0.15
0.10
0.05
0.00
January
February
March
April
May
June
July
August
September
October
November
December

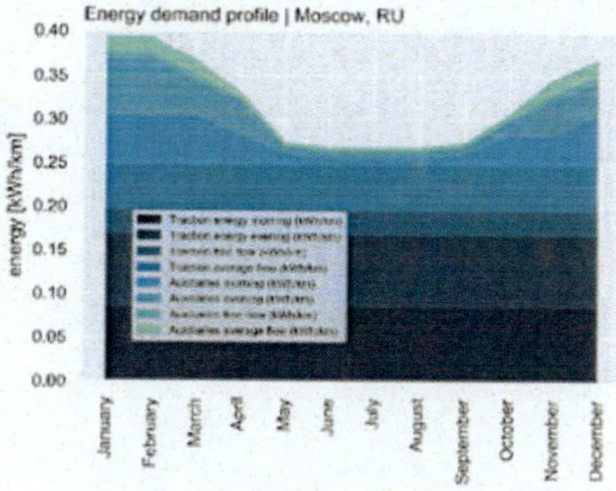
Energy demand profile | Moscow, RU
energy [kWh/km]
0.40
0.35
0.30
0.25
0.20
0.15
0.10
0.05
0.00
January
February
March
April
May
June
July
August
September
October
November
December

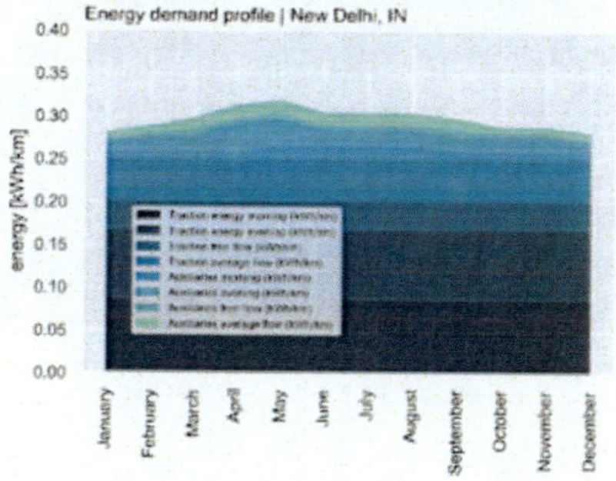
Energy demand profile | New Delhi, IN
energy [kWh/km]
0.40
0.35
0.30
0.25
0.20
0.15
0.10
0.05
0.00
January
February
March
April
May
June
July
August
September
October
November
December

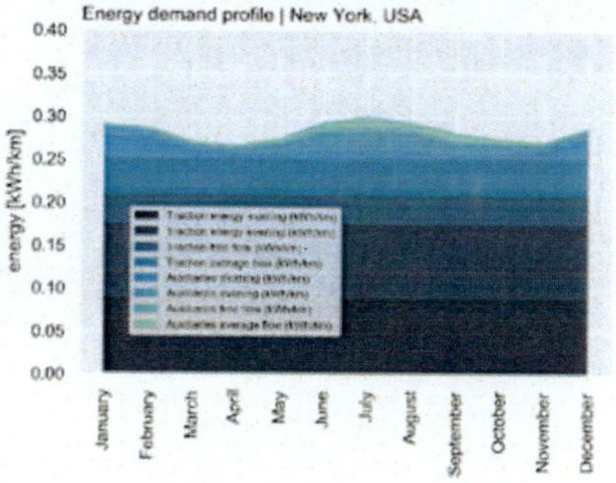
Energy demand profile | New York, USA
energy [kWh/km]
0.40
0.35
0.30
0.25
0.20
0.15
0.10
0.05
0.00
January
February
March
April
May
June
July
August
September
October
November
December

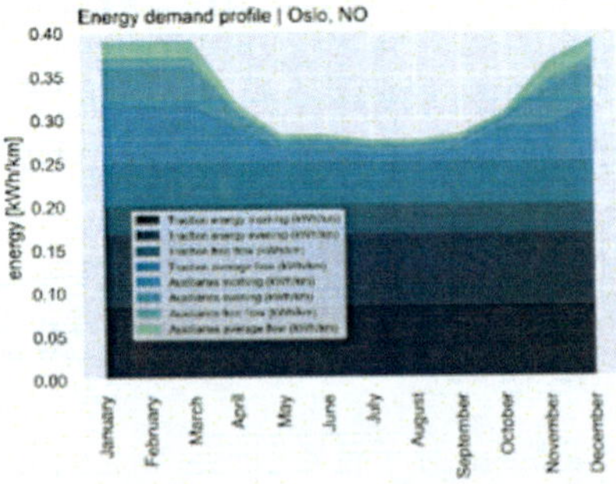
Energy demand profile | Oslo, NO
energy [kWh/km]
0.00 0.05 0.10 0.15 0.20 0.25 0.30 0.35 0.40
January February March April May June July August September October November December

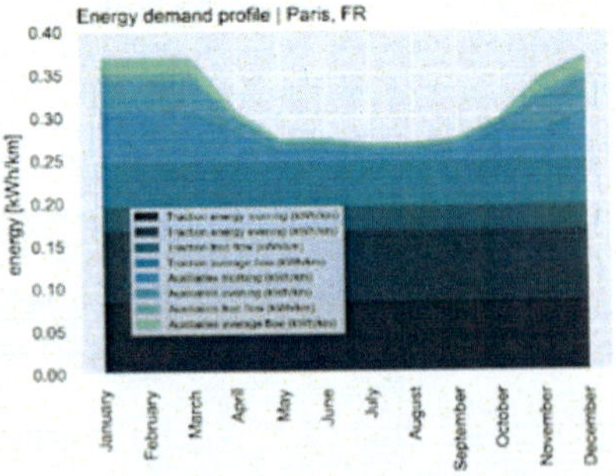
Energy demand profile | Paris, FR
energy [kWh/km]
0.00 0.05 0.10 0.15 0.20 0.25 0.30 0.35 0.40
January February March April May June July August September October November December

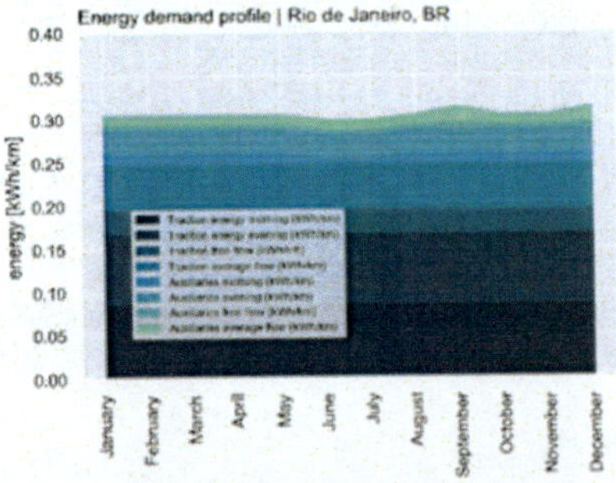
Energy demand profile | Rio de Janeiro, BR
energy [kWh/km]
0.00 0.05 0.10 0.15 0.20 0.25 0.30 0.35 0.40
January February March April May June July August September October November December

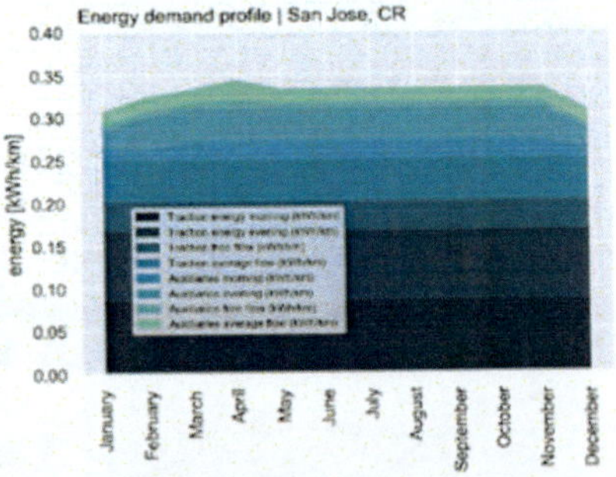
Energy demand profile | San Jose, CR
energy [kWh/km]
0.00 0.05 0.10 0.15 0.20 0.25 0.30 0.35 0.40
January February March April May June July August September October November December

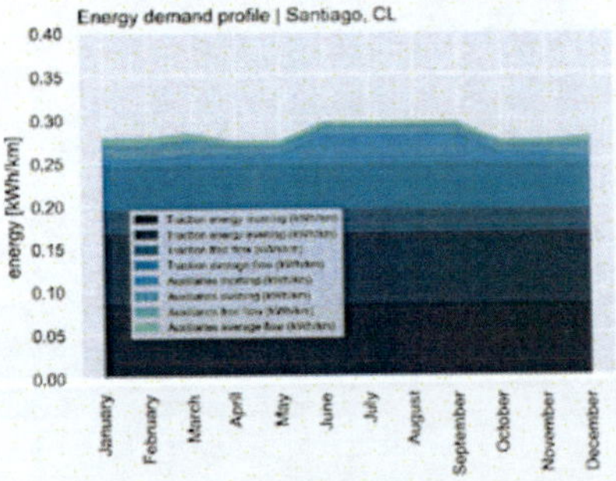
Energy demand profile | Santiago, CL
energy [kWh/km]
0.00 0.05 0.10 0.15 0.20 0.25 0.30 0.35 0.40
January February March April May June July August September October November December

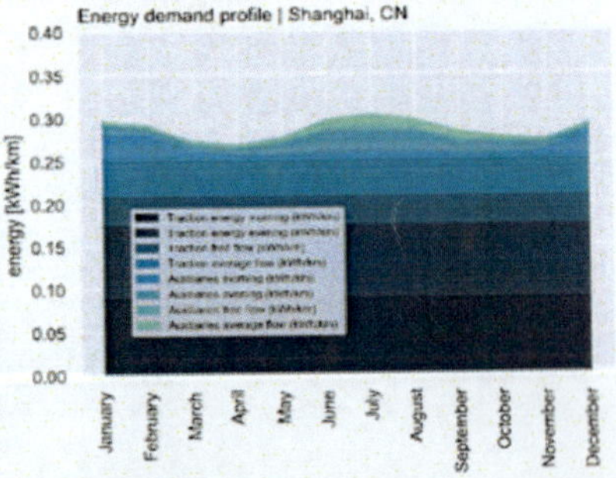
Energy demand profile | Shanghai, CN
energy [kWh/km]
0.00 0.05 0.10 0.15 0.20 0.25 0.30 0.35 0.40
January February March April May June July August September October November December

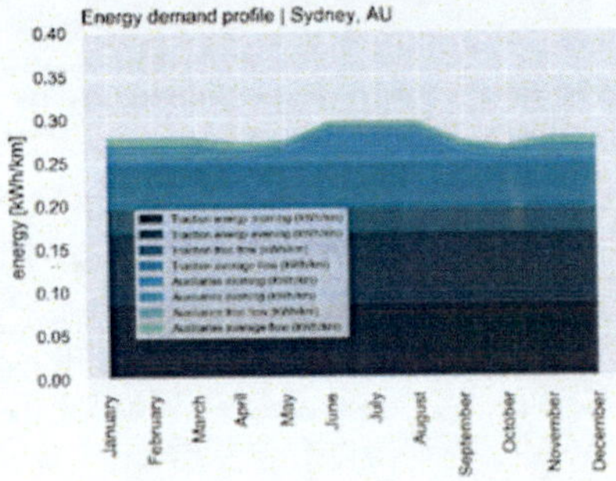
Energy demand profile | Sydney, AU
energy [kWh/km]
0.00 0.05 0.10 0.15 0.20 0.25 0.30 0.35 0.40
January February March April May June July August September October November December

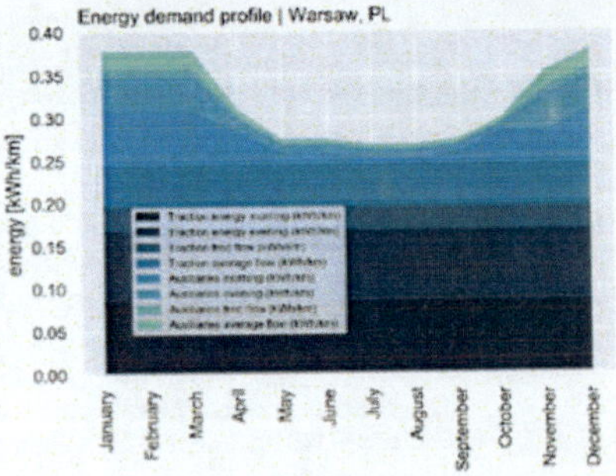
Energy demand profile | Warsaw, PL
energy [kWh/km]
0.00 0.05 0.10 0.15 0.20 0.25 0.30 0.35 0.40
January February March April May June July August September October November December

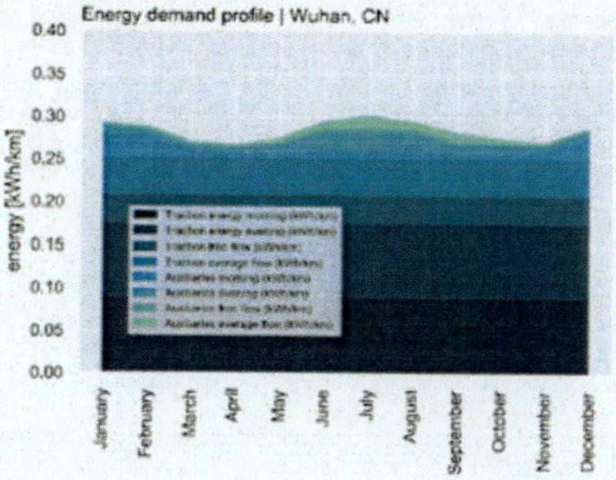

G. Break even analysis for all product systems created in all 25 cities considered in the usage stage.

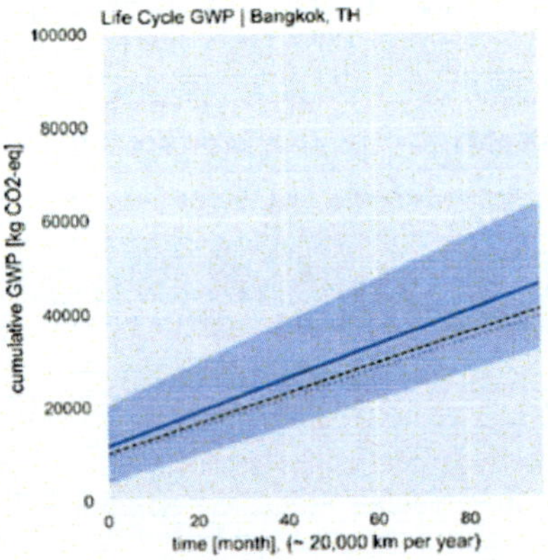
Life Cycle GWP | Bangkok, TH
cumulative GWP [kg CO2-eq]
0
20000
40000
60000
80000
100000
0
20
40
60
80
time [month], (~ 20,000 km per year)

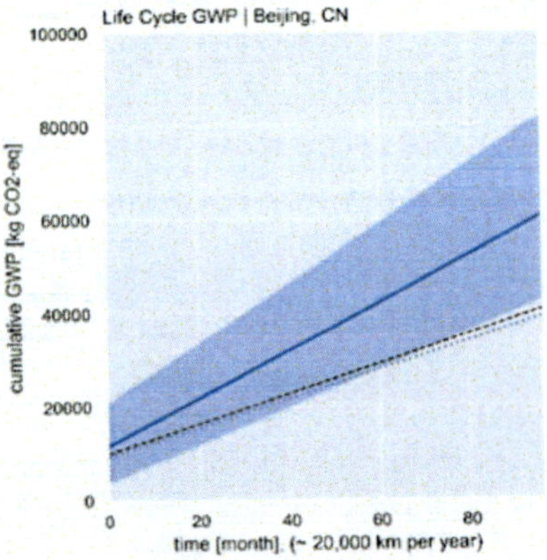
Life Cycle GWP | Beijing, CN
cumulative GWP [kg CO2-eq]
0
20000
40000
60000
80000
100000
0
20
40
60
80
time [month], (~ 20,000 km per year)

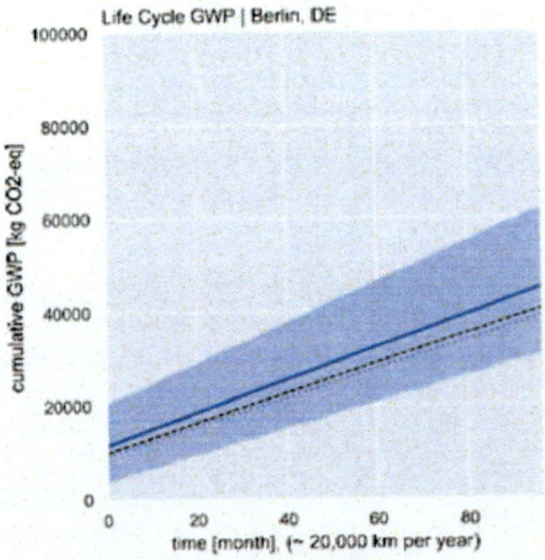
Life Cycle GWP | Berlin, DE
cumulative GWP [kg CO2-eq]
0
20000
40000
60000
80000
100000
0
20
40
60
80
time [month], (~ 20,000 km per year)

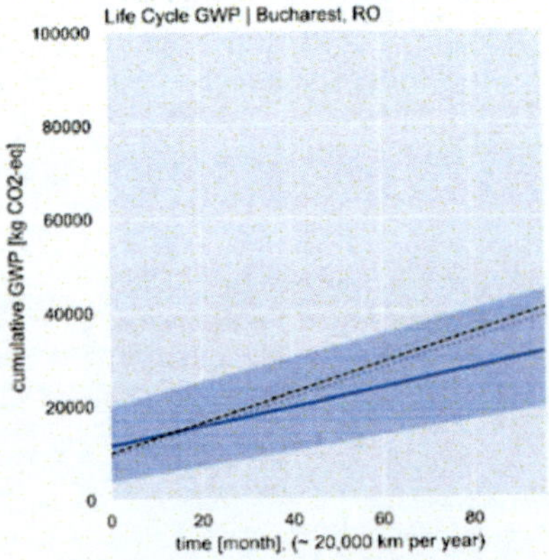
Life Cycle GWP | Bucharest, RO
cumulative GWP [kg CO2-eq]
0
20000
40000
60000
80000
100000
0
20
40
60
80
time [month], (~ 20,000 km per year)

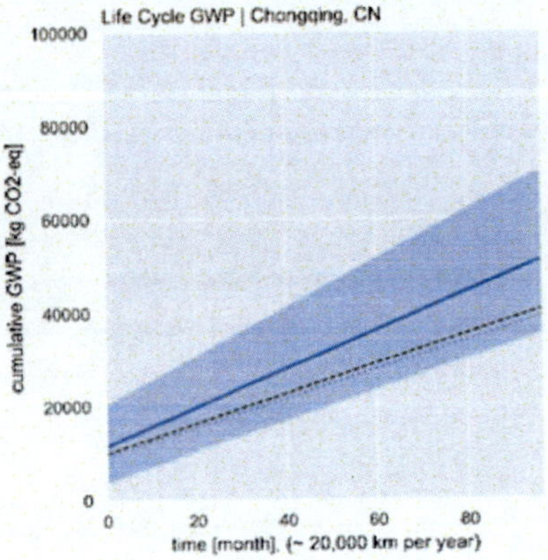
Life Cycle GWP | Chongqing, CN
cumulative GWP [kg CO2-eq]
0
20000
40000
60000
80000
100000
0
20
40
60
80
time [month], (~ 20,000 km per year)

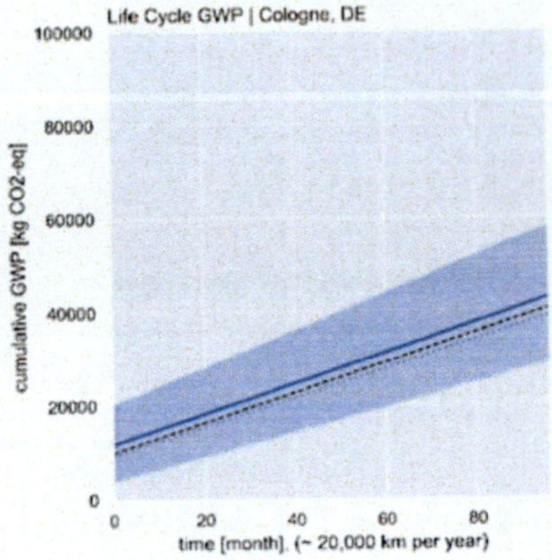
Life Cycle GWP | Cologne, DE
cumulative GWP [kg CO2-eq]
0
20000
40000
60000
80000
100000
0
20
40
60
80
time [month], (~ 20,000 km per year)

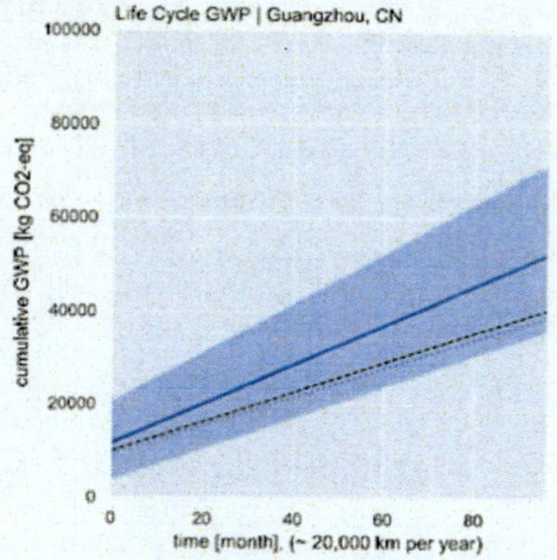
Life Cycle GWP | Guangzhou, CN
cumulative GWP [kg CO2-eq]
0
20000
40000
60000
80000
100000
0
20
40
60
80
time [month], (~ 20,000 km per year)

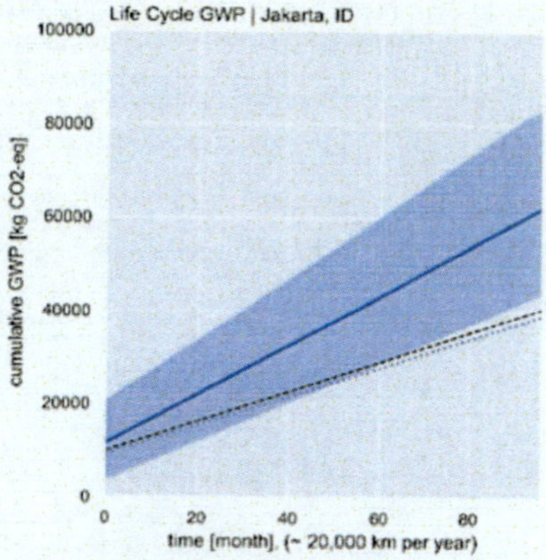
Life Cycle GWP | Jakarta, ID
cumulative GWP [kg CO2-eq]
0
20000
40000
60000
80000
100000
0
20
40
60
80
time [month], (~ 20,000 km per year)

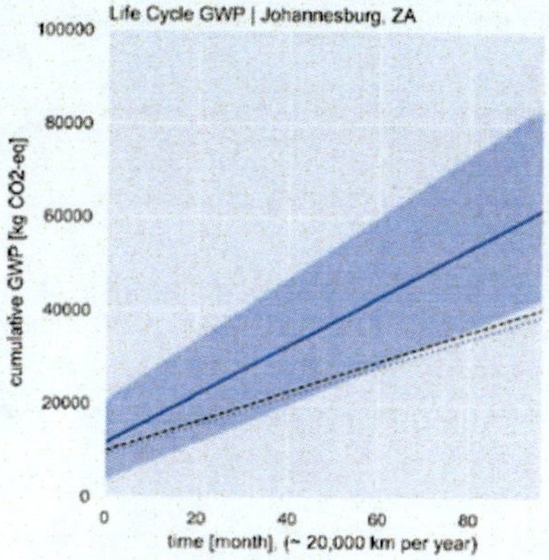
Life Cycle GWP | Johannesburg, ZA
cumulative GWP [kg CO2-eq]
0
20000
40000
60000
80000
100000
0
20
40
60
80
time [month], (~ 20,000 km per year)

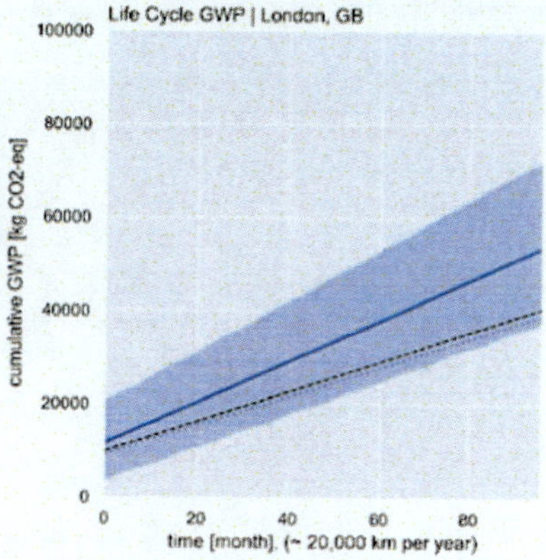
Life Cycle GWP | London, GB
cumulative GWP [kg CO2-eq]
0
20000
40000
60000
80000
100000
0
20
40
60
80
time [month], (~ 20,000 km per year)

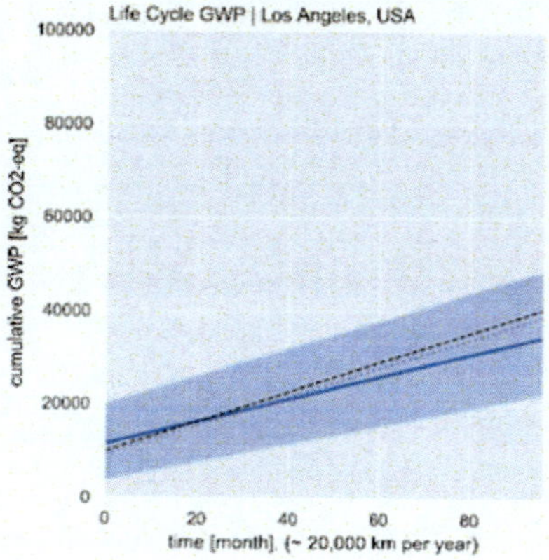
Life Cycle GWP | Los Angeles, USA
cumulative GWP [kg CO2-eq]
0
20000
40000
60000
80000
100000
0
20
40
60
80
time [month], (~ 20,000 km per year)

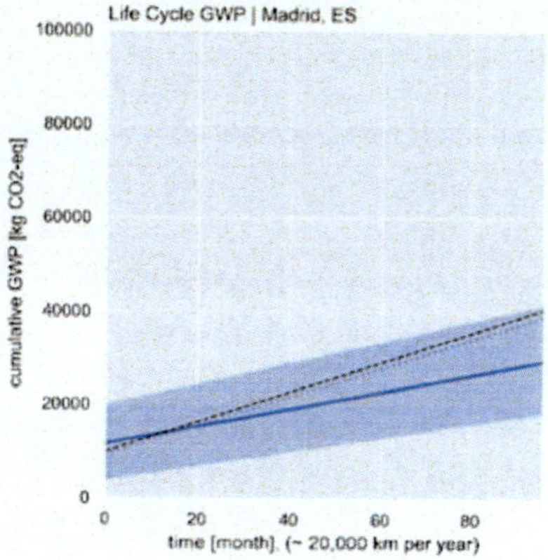
Life Cycle GWP | Madrid, ES
cumulative GWP [kg CO2-eq]
0
20000
40000
60000
80000
100000
0
20
40
60
80
time [month], (~ 20,000 km per year)

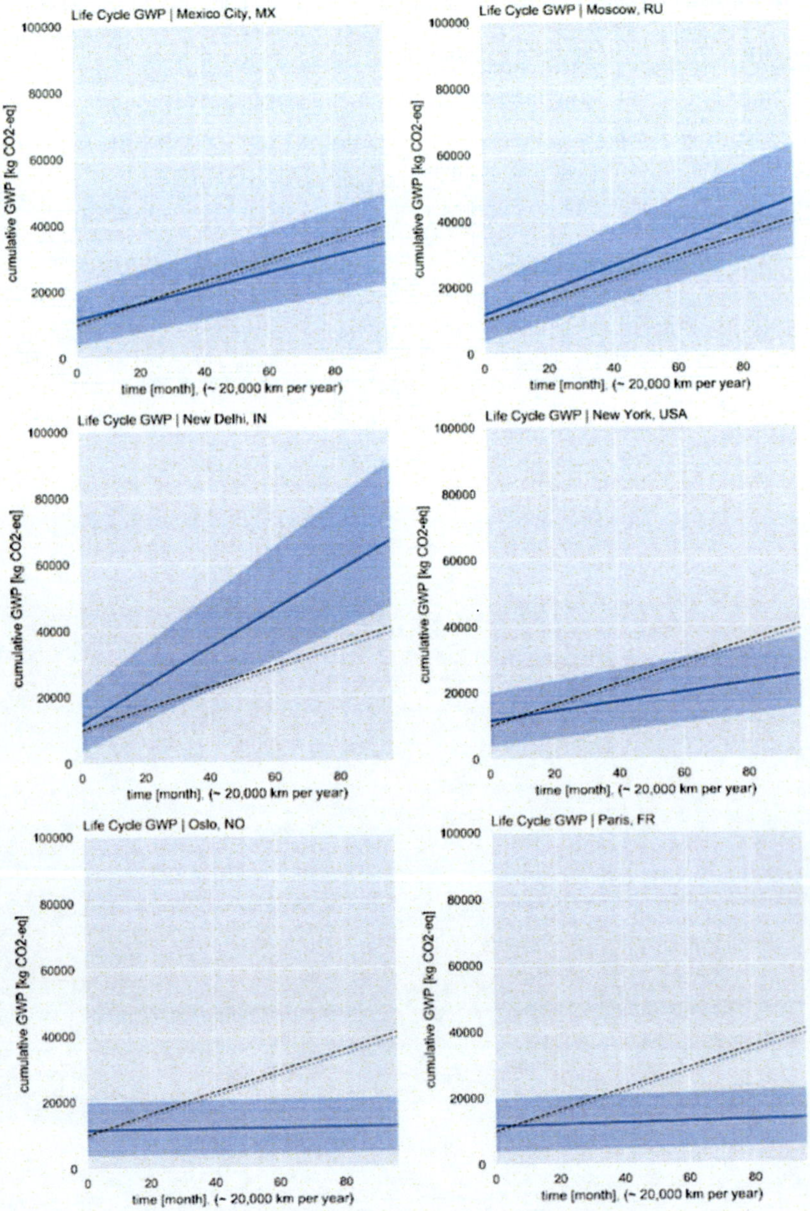

Life Cycle GWP | Mexico City, MX
Life Cycle GWP | Moscow, RU
Life Cycle GWP | New Delhi, IN
Life Cycle GWP | New York, USA
Life Cycle GWP | Oslo, NO
Life Cycle GWP | Paris, FR
cumulative GWP [kg CO2-eq]
time [month], (~ 20,000 km per year)
0
20000
40000
60000
80000
100000
0
20
40
60
80

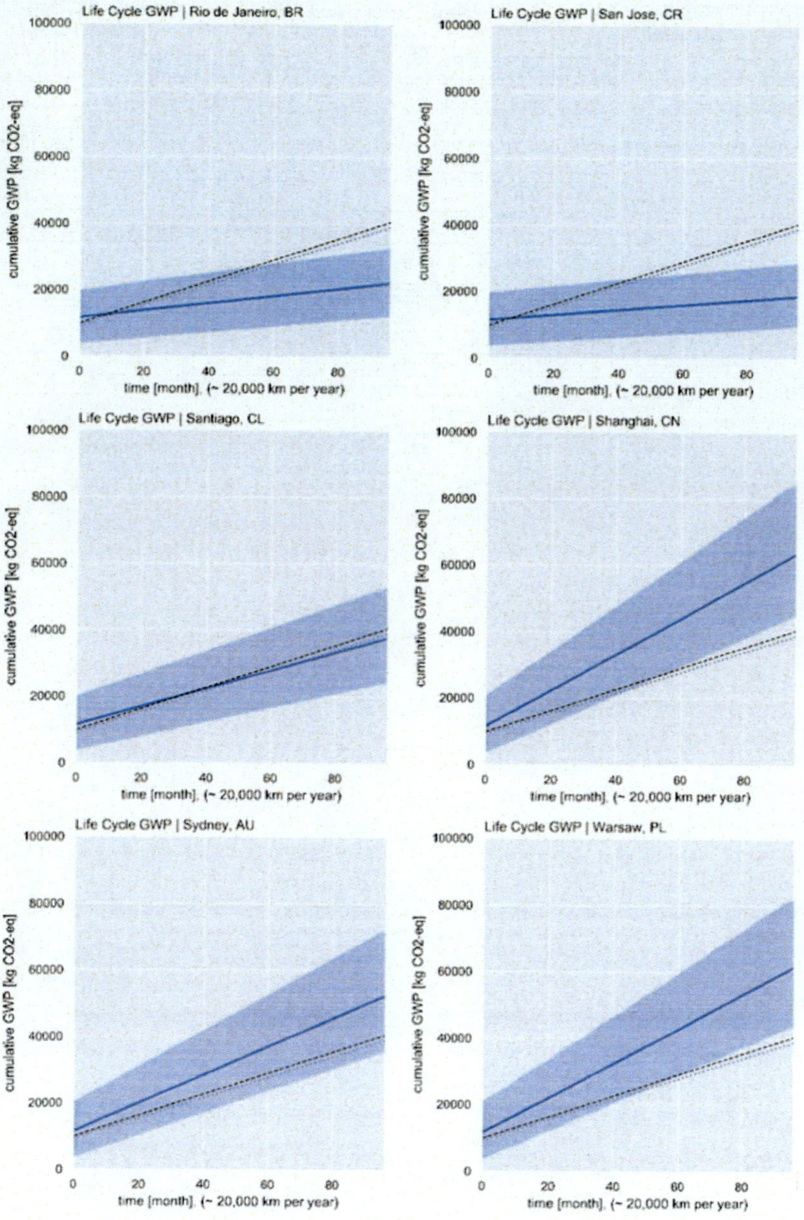
Life Cycle GWP | Rio de Janeiro, BR
Life Cycle GWP | San Jose, CR
Life Cycle GWP | Santiago, CL
Life Cycle GWP | Shanghai, CN
Life Cycle GWP | Sydney, AU
Life Cycle GWP | Warsaw, PL
cumulative GWP [kg CO2-eq]
100000
80000
60000
40000
20000
0
time [month], (~ 20,000 km per year)
0
20
40
60
80

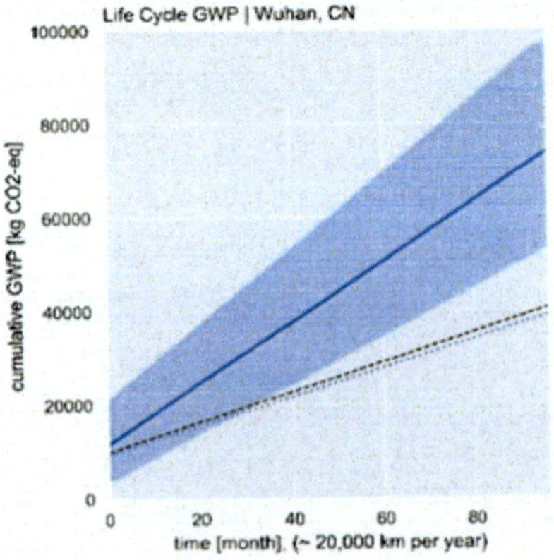
Life Cycle GWP | Wuhan, CN
cumulative GWP [kg CO2-eq]
100000
80000
60000
40000
20000
0
0
20
40
60
80
time [month], (~ 20,000 km per year)